Switchable Bioelectronics

Switchable Bioelectronics

edited by
Onur Parlak

Published by

Jenny Stanford Publishing Pte. Ltd.
Level 34, Centennial Tower
3 Temasek Avenue
Singapore 039190

Email: editorial@jennystanford.com
Web: www.jennystanford.com

British Library Cataloguing-in-Publication Data
A catalogue record for this book is available from the British Library.

Switchable Bioelectronics

ISBN 978-981-4800-89-1 (Hardcover)
ISBN 978-1-003-05600-3 (eBook)

Contents

Preface

I am very delighted to introduce this very first book on switchable bioelectronics to serve the scientific community with topical critical and tutorial reviews covering different aspects of polymer chemistry and engineering technologies.

In this book, we aim to give a collective summary and several applications of the rapidly emerging field of switchable interfaces and its implications for bioelectronics. We bring various aspects of the field together, represent the early breakthroughs and key developments, and highlight and discuss the future of switchable bioelectronics by focusing on bioelectrochemical processes based on mimicking and controlling biological environments with external stimuli. All these studies strive to answer the fundamental question How do living systems probe and respond to their surroundings? And, following on from that, how can one transform these concepts to serve the practical world of bioelectronics? The central obstacle to this vision is the absence of versatile interfaces that are able to control and regulate the means of communication between biological and electronic systems.

Here, all chapters focus on the overall progress made to date in building such interfaces at the level of individual biomolecules and highlight the latest efforts to generate device platforms that integrate biointerfaces with electronics. In Chapter 1, Parlak introduces the general concept of dynamic interfaces for bioelectronics and give an overview of the importance of materials and systems for switchable bioelectronics, introducing the reader to different biointerfaces. In Chapter 2, Beyazit pieces together different types of stimuli-responsive polymers and applications. In Chapter 3, Ozaydin-Ince et al. lay special emphasis on stimuli-responsive polymers with tunable release kinetics and describe the importance of polymer design for delivery applications. Moving towards applying these principles to develop innovative devices for bioelectronic applications, Yildiz et al., in Chapter 4, review the field of conformational switching in nanofibers for gas sensing applications. Finally, in Chapter 5, Basan and Yılmaz focus on molecular imprinting polymers as recognition elements for sensing applications. Each of these chapters concludes with a discussion of key examples in their respective areas and their implications in the field of switchable bioelectronics.

I hope you will enjoy going through this book as much as I have enjoyed putting it together. I thank all the contributing authors for their enthusiasm and patience, and the editorial team of Jenny Stanford Publishing for all the diligent work enabling the publication of this book.

Onur Parlak

Chapter 1

Introduction to Dynamic Bioelectronic Interfaces

Onur Parlak
Department of Materials Science and Engineering, Stanford University, Stanford, CA 94305, USA
parlak@stanford.edu

Dynamic interfaces with the ability to respond to physical and chemical environmental changes have been employed in many different research fields, from environmental studies to bioelectronics.[1] Having the ability to control and regulate interfacial properties is essential in most of these studies.[2] Over the years, many different materials and approaches have been developed to understand the interfacial properties of dynamic interfaces (Fig. 1.1).[1] In this chapter, the general concepts of dynamic interfaces and characterization methods are discussed and demonstrated. The emerging field of dynamic biointerfaces and their applications in bioelectronics is extensively reviewed.

Switchable Bioelectronics
Edited by Onur Parlak

ISBN 978-981-4800-89-1 (Hardcover), 978-1-003-05600-3 (eBook)
www.jennystanford.com

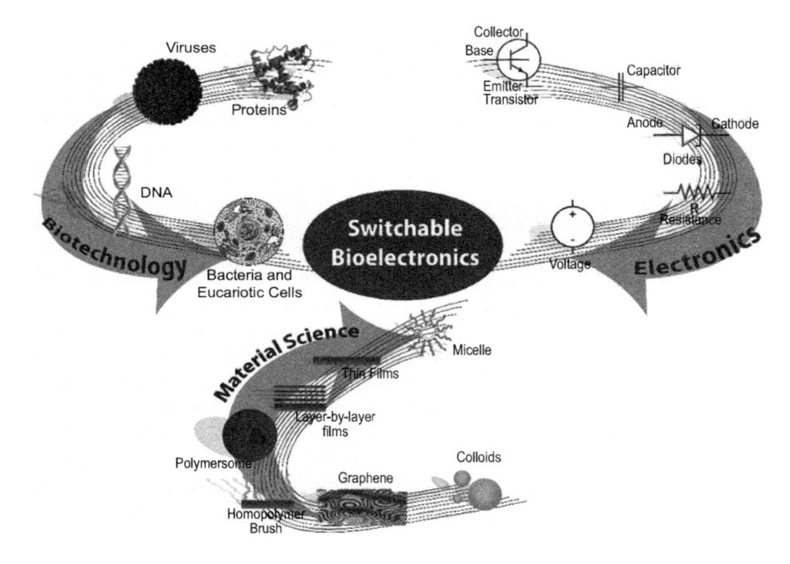

Figure 1.1 Schematic illustration of how biotechnology, electronics, and materials science merge to form the concept of bioelectronics. Reprinted with permission from Ref. [1] Copyright © 2015 Elsevier B.V.

1.1 Switchable Biointerfaces

Dynamic biointerfaces are considered to be systems that show a responsive behavior and the ability to adapt to physical and/or chemical environmental changes in their surroundings.[3] Also, dynamic interfaces are able to show on-demand microscopic and interfacial changes upon interaction with their environments.[1,3] These changes are mainly categorized into physical and chemical changes. Physical changes include temperature fluctuations, light intensity differences, and change in the strength of magnetic or electrical fields.[1] Chemical changes include change in the pH and ionic strengths and the addition of different chemicals to reaction media.[1,3]

There are two main strategies to construct dynamic interfaces, polymeric films and self-assembly monolayer (SAM) formations.[4] Using the SAM approach, self-organized and adaptive interfaces can be generated. The SAM approach allows one to tune the thickness, structure, and surface energy of a responsive interface by introducing different types of monolayers. Even though the SAM technology was developed a long time ago, switchable SAMs have been studied only recently to overcome the static features of conventional SAMs.[4]

Another common strategy to generate dynamic interfaces is to incorporate stimuli-responsive polymers.[5–7] The general methods to form responsive polymeric films on solid supports are spin-coating, chemical vapor deposition, laser ablation, and chemical and/or electrochemical approaches.[1,4] Despite the fact that spin-coating is considered a relatively simple and practical method to generate polymeric films on solid surfaces, it suffers from some technical difficulties, such as the production of a large amount of waste material and rather thin films when less soluble polymers are used.[5] Other techniques, such as laser ablation and vapor deposition, are less practical. The electrochemical technique is relatively easier; however, problems and mismatches may result after the formation of the polymer film on the surface due to the incompatibility of the polymer film with the electrolytes and sensitivity to different electrochemical conditions, such as temperature and electrical stimuli.[1] It is also impossible to grow polymeric films on an insulating surface via electrochemical polymerization. One straightforward method to grow robust and controlled polymeric films on solid supports is to use surface-tethered polymers, also known as polymer brushes.[8] Polymer brushes are constructed by tethering of individual long-chain polymer molecules to specific anchor sites on the surface. In this way, a densely packed polymeric surface can be obtained and the random-walk conformation/motion of the polymer molecules that can occur in the polymer solution or the solution-casted polymer can be restricted. Polymer brushes can be obtained either by physical or chemical methods, but chemical covalent linkage is preferred due to the stability it provides. There are two main approaches available for forming polymer brushes on a solid surface.

In the "grafting to" approach, externally synthesized polymer molecules with different molecular weights are attached to the previously functionalized surface. However, in the "grafting from" technique, or the so-called surface-initiated polymerization, the polymer is grown from the surface via attachment of monomer units to the prefunctionalized surface.[9] The grafting-from approach provides more advantages compared to its grafting-to counterpart. In the grafting-from approach, a denser surface can be obtained because the grafting-to approach usually suffers from steric hindrance after initial grafting of the polymer chains. Surface-initiated polymerization can be performed using many different polymerization methods, including free-radical, anionic, cationic, atom-transfer radical, ring-opening metathesis, reversible addition-fragmentation transfer, and nitroxide-mediated radical polymerization.[9]

1.2 Physically Stimulated Systems

1.2.1 Light-Switchable Interfaces

One of the most common stimuli to obtain dynamic interfaces is light.[10] There are many different types of light-responsive molecules/materials, including azobenzenes, spiropyrans (SPs), and azulenes, and many in the form of molecules, polymers, SAMs, and conjugated nanoparticles. Each of these light-responsive systems has a unique switching mechanism, isomerization rate, and intensity and the system parameters are heavily affected by these parameters.[11]

In one of our previous studies, we developed a light-responsive dynamic biointerface mainly composed of a light-responsive methacrylated SP molecular unit copolymerized with polyacrylamide (PAAM).[5] We used this system to control and regulate enzyme-based molecular interaction. Using electrochemical measurements, we showed that interfacial bioelectrochemical properties can be regulated by changing the wavelength of light (Fig. 1.2) The system depends on designing a positively responding biointerface using a covalently attached polymer containing a light-responsive polyacrylamide polymer copolymerized with SP molecular units and then conjugated with the graphene on an electrode. In the polymer chain, SP units have two isomeric states, (i) charge-separated merocyanine under ultraviolet (UV) light conditions and (ii) SP under visible light or dark conditions. The isomeric changes result in two important features of the interface—structural permeability and polarity differences. Here, these two states are reversibly controlled upon alternating UV and

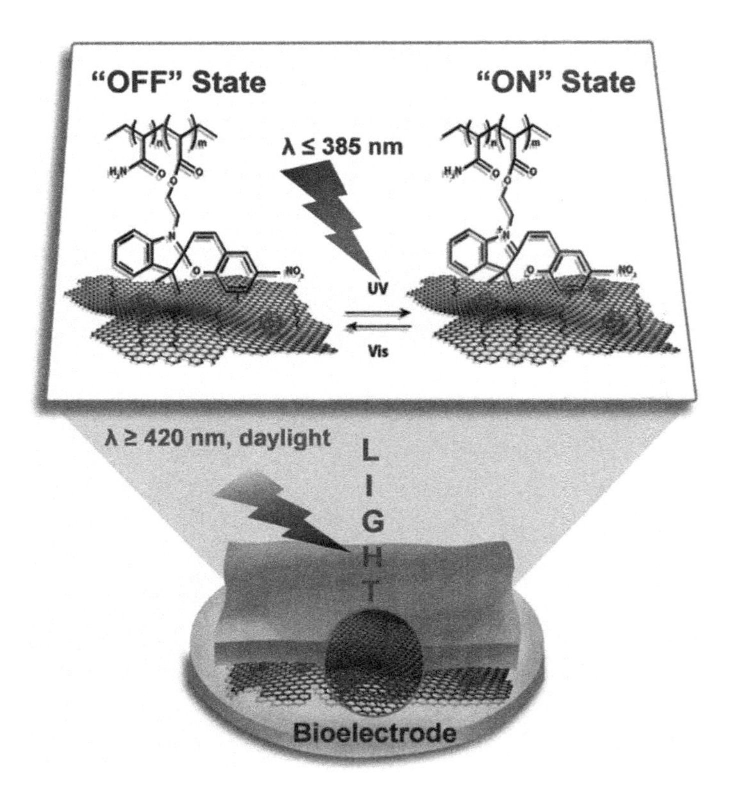

Figure 1.2 Schematic of the light-switchable bioelectrocatalytic graphene interface. Reprinted with permission from Ref. [5] Copyright © 2015 WILEY-VCH Verlag GmbH & Co. KGaA, Weinheim.

visible light. When the system is irradiated with UV light (≤385 nm), SP undergoes a ring-opening conformation. This isomerization results in volume and polarity changes. These changes are responsible for inducing permeability in the polymer structure (i.e., volume effect) and an increase in the conductivity (i.e., polarity effect). However, when the system is irradiated with light of a higher wavelength (≥420 nm), the polymer structure turns into

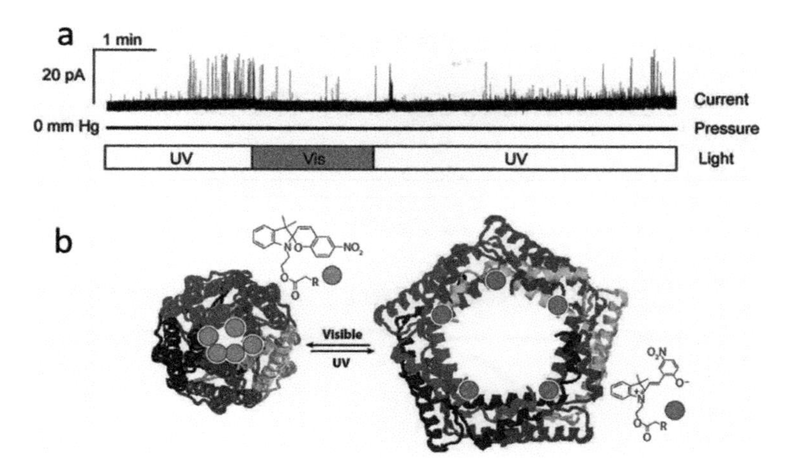

Figure 1.3 Electrophysiological analysis of the reversible functioning of modified MscL by the patch-clamp technique (a) and illustration of a switchable biohybrid nanovalve based on the mechanosensitive channel protein of large conductance together with spiropyran molecular units (b). Reprinted with permission from Ref. [1] Copyright © 2015 Elsevier B.V.

a densely packed form at the electrode interface, which results in lower permeability and diffusion of the substrate, suppressing the electrochemical signals. As a result, the substrate cannot freely access the surface.[5]

Another way to incorporate SP-based light-responsive molecules is to use them in a mechanosensitive channel protein of large conductance (MscL from *Escherichia coli*) as the molecular valve to control channel transport channel in the cells (Fig. 1.3).[10] In many cells, MscL is used as a safety valve to control osmotic downshifts in bacterial cells or instant influx of water to prevent high tension in the membrane. They can be opened to around 3 mm to allow the flux of ions and proteins to prevent cell lysis.[10]

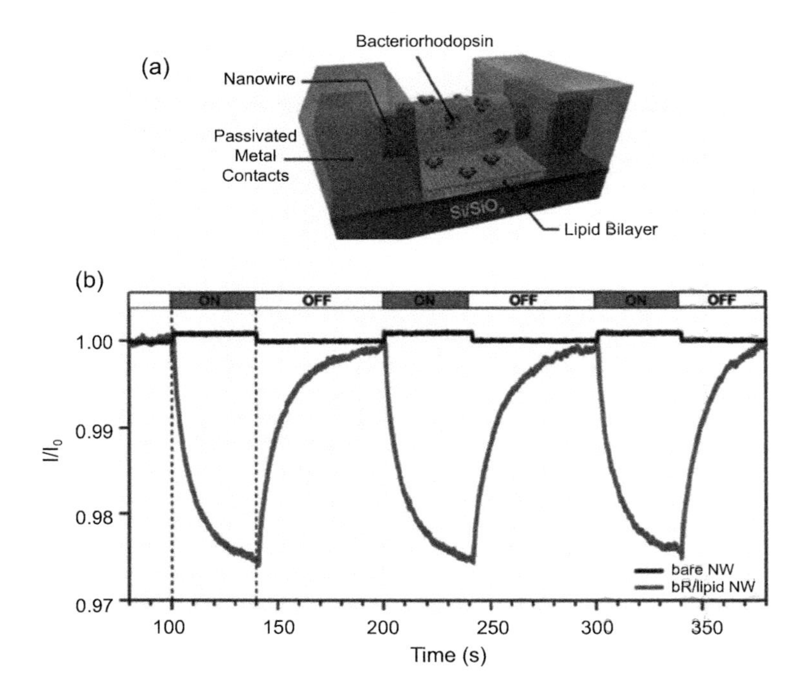

Figure 1.4 Schematic illustration of (a) a light-activated Si NW–based FET device containing a nanowire covered with a lipid bilayer and bacteriorhodopsin protein and (b) channel current analysis of a SiNW-based FET device. Reprinted with permission from Ref. [1] Copyright © 2015 Elsevier B.V.

In this study, researchers used the same strategy to follow single-channel recording at 20 mV without any pressure gradient, and the channel shows an upward current. However, when the channel is illuminated with UV light, the channel pressure and current change significantly. This research opens up a new route to externally controlled ion pumps for various applications.[10]

In another study, researchers designed a light-driven bioelectronic device consisting of a 1D lipid bilayer on a

Si-nanowire-based field-effect transistor (Fig. 1.4a).[12] In this particular device, they combined bacteriorhodopsin protein with a lipid bilayer and coated it on a nanowire surface. They tested the device behavior under two different light conditions and obtained different channel currents based on the absorbance of light by the protein resulting in a multistate photocycle that moves the proton across the membrane. They were able to regulate the current and gain back the signal without any significant loss (Fig. 1.4b).[12] On the basis of the same principle, another group demonstrated a similar approach using a photoreceptor protein and developed light-powered biocapacitor and were also able to regulate device behavior in two different states.[13]

1.2.2 Temperature-Switchable Interfaces

One of the other effective ways to produce a dynamic interface is to use temperature.[1,4] The fabrication of SAMs or polymeric films in the form of a polymer brush or hydrogel is a common approach to generating stable films on a surface.

Temperature-based responsive interfaces are mainly based on the mechanisms of reversible control of molecular recognition events, cell attachment to a solid surface, and enzyme-based biocatalysis.[1] The most common material to generate a responsive interface is poly(*N*-isopropylacrylamide [PNIPAAM]). In our recent study, we used PNIPAAM with a graphene-cholesterol oxidase hybrid to develop a temperature-responsive, switchable, positively responding biointerface.[7] The system consists of a graphene as a donor and PNIPAAM as a receptor, assembled together

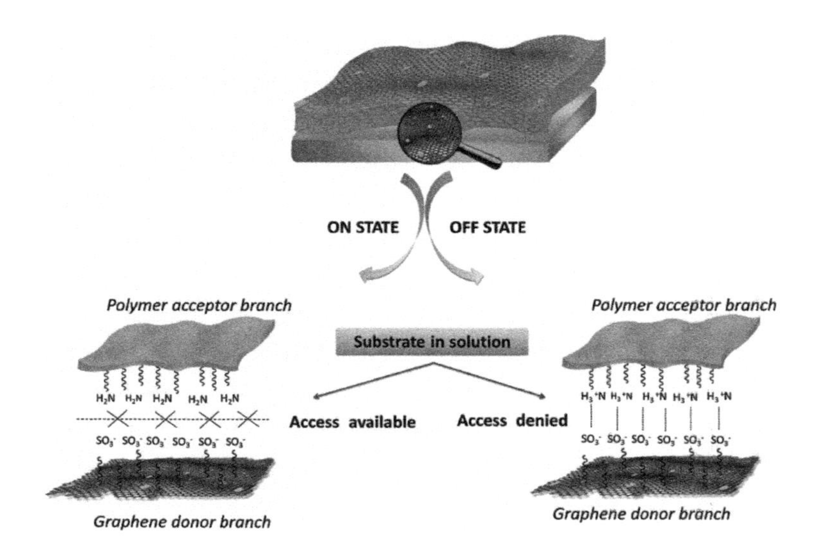

Figure 1.5 Schematic illustration of temperature-responsive bioelectrocatalytic graphene interfaces in two different states. Reprinted with permission from Ref. [1] Copyright © 2015 Elsevier B.V.

on the basis of the donor-receptor interaction. At a low temperature (i.e., at 20°C), hydrogen bonding interaction creates a coalescence on the surface, causing considerable shrinkage in the donor-to-receptor interface. In this way, enzyme-substrate interaction is restricted and this results in a decrease in the diffusion of reactants and the consequent activity of the system. However, at high temperatures, such as 40°C, the donor-receptor interaction is diminished, making the substrate easier to access to facilitate bioelectrocatalysis (Fig. 1.5).[7]

One of the significant advantages of using PNIPAAM for bioelectronic applications is to have a sharp difference in solution properties near body temperature. This makes it possible to change the conditions and the physiological state of the biomolecules as desired. On the basis of this concept, researchers developed programmable adsorption and release of proteins using a microfluidic channel. They modified a microfluidic channel with PNIPAAM about 4 nm thick by combining gold heater wires with Si_3N_4 membranes, as shown in Fig. 1.6a.[14] This combination makes it possible to heat and cool each gold wire specifically to control and release biomolecules, including myoglobin, cytochrome C, and bovine serum albumin (BSA), in seconds.

1.2.3 Electrically Switchable Interfaces

Another effective and noninvasive way to form a dynamic interface is to use an electrical field to control and regulate interfacial properties in the area of bioelectrocatalysis, prevention of marine biofouling and electrochromic windows applications.[16]. The most commonly employed method to generate a dynamic interface is to use SAMs instead of a polymeric structure due to the lack of or a lower rate of conformational change of polymers.[16]

In one of the earliest examples, researchers used electrical potential as an external stimulus to regulate cell attachment during cell growth or reversible adsorption and release of protein onto electrically responsive structure–modified solid substrates. Researchers have successfully shown an electrically responsive interface consisting of a SAM to control the early stages of bacterial adhesion. The mechanism is based on controlling the charge of the end group in the layer with applied potential (Fig. 1.7a).[15]

In this study, researchers tested the adhesive behavior of a bacterium under two different applied potential conditions and compared their results with an open-circuit

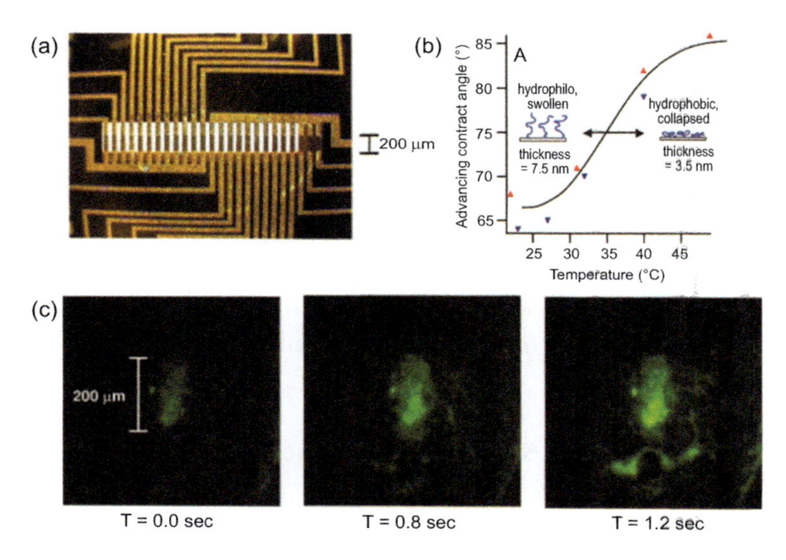

Figure 1.6 A photomicrograph of a microfluidic chip showing (a) an array of gold heater tracks on top of a 200 μm wide Si_3N_4 membrane (white) array, (b) the water-contact angle measurements of PNIPAAM films as a function of temperature, and (c) fluorescence microscopy images of fluorescein-labeled myoglobin (green) interacting with a single heated track. The image obtained on heating a track above the PNIPAAM transition temperature after exposing it to a 0.5 mg/mL myoglobin solution followed by rinsing it in myoglobin-free buffer. The images were obtained 0.8 and 1.2 s after the hot track was turned off, releasing a plume of protein into a stagnant solution. Reprinted with permission from Ref. [1] Copyright © 2015 Elsevier B.V.

voltage. The confocal images in Fig. 1.7b show that bacterial adhesion was successfully controlled under two different states.[15]

In another study, researchers developed composite materials mainly composed of copper and polyacrylic acid (PAA) and modified their electrode with this copper-polymer composite, which shows switchable redox activity

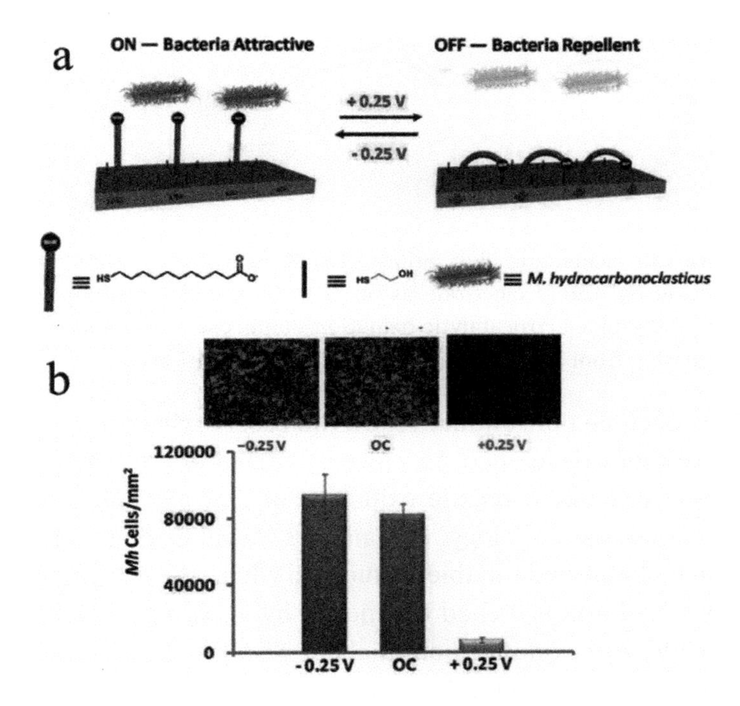

Figure 1.7 Schematic representation of (a) switchable adhesion of bacterial cells on the basis of charge and conformational change of surface-modified groups upon applied potential. The conformation of the surface molecule is changed on applying potential, and adhesion of bacterial cells is controlled; (b) confocal microscopy images and cell counting graph in three different applied potential conditions: −0.25 V, open-circuit (OC), and +0.25 V. Reprinted with permission from Ref. [1] Copyright © 2015 Elsevier B.V.

depending on the location of Cu ions/particles inside the polymeric network.[17] When the copper is close to the electrode surface, the system shows metallic conductivity. However, if the position of the copper ions/particles changes, the oxidation state of the copper changes from nanoparticle (Cu^0) form to ionic form (Cu^{2+}) and the

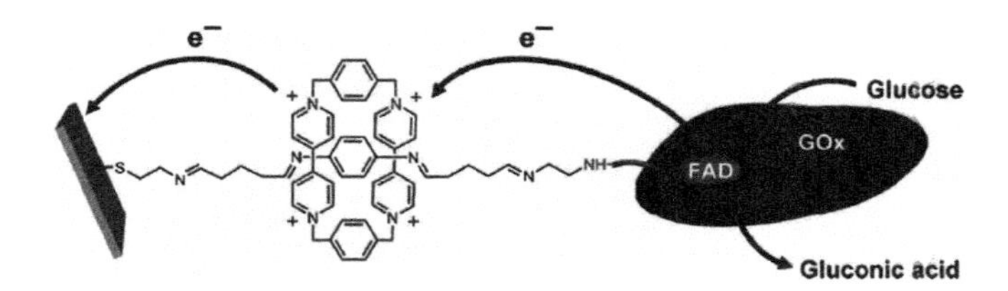

Figure 1.8 Schematic illustrations of electrochemically stimulated rotaxane on a gold electrode as an electron-transfer mediator to control the bioelectrocatalytic oxidation of glucose. Reprinted with permission from Ref. [1] Copyright © 2015 Elsevier B.V.

films become nonconductive. On the basis of this principle, researchers developed an electroswitchable and tunable biofuel cell based on the oxidation of glucose. Following the same methodology, researchers developed an electroswitchable and tunable biofuel cell where the biofuel cell was tuned and switched on and off by applying reductive (−0.5 V) and oxidative potentials (+0.5 V) to the electrode surface, which reversibly yielded the conductive Cu^0-PAA (on state) or the nonconductive Cu^{2+}-PAA (off state) forms of the polymeric film. Using the same approach, researchers also successfully wired the enzyme molecules and enzyme cofactor to the electrode surface using the electrochemically active molecule rotaxane. In this study, rotaxane acted as a redox relay by mediating electron transfer on application of a potential. This approach makes it possible to develop tunable/controllable bioelectronic devices (Fig. 1.8).[18]

1.2.4 Magnetoswitchable Interfaces

One of the first methods to generate a dynamic interface was to use magnetic materials.[17] The common way to construct a magnetically switchable interface is by employing

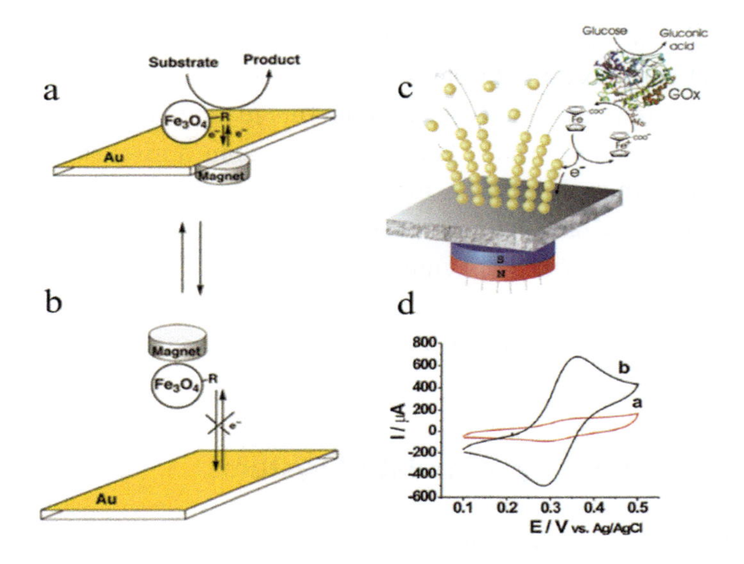

Figure 1.9 Schematic illustrations of (a, b) a reversible electrocatalytic reaction with functional magnetic nanoparticles (Fe_3O_4-R) controlled by an external magnetic field; (c) schematics of magnetically controlled, reversible assembly of gold-coated magnetic nanoparticles on an electrode surface; and (d) response of mediated electrochemical oxidation of glucose in the presence (black curve) and absence (red curve) of a magnetic field. Reprinted with permission from Ref. [1] Copyright © 2015 Elsevier B.V.

magnetic nanoparticles of different sizes, shapes, and surface functionalizations. Magnetic nanoparticles are incorporated into and/or modified with redox-active materials and usually used as switchable electrochemical mediators in the area of bioelectrocatalysis.[18] Following this approach, researchers have recently developed various power-on-demand biofuel cells and biosensor devices, as described in Fig. 1.9.[11]

In this study, redox-modified nanoparticles are controlled with an external magnet, which can activate mediated electron transfer between the electrical contact and the redox unit. Researchers were also able to deactivate the system by simply switching the magnet away from the electrode surface. Following the same approach, the same group developed magnetically controlled bioelectrocatalysis based on glucose oxidation in the presence of glucose oxidase enzyme (Fig. 1.9c,d).[11] This approach allows controlling and regulating biocatalytic reactions using redox-active species and magnetic nanoparticles.

1.3 Chemically Stimulated Systems

In addition to physically stimulated responsive interfaces, another general strategy is to use chemical stimuli. Chemical control of the interface is usually provided by addition or in situ generation of some (bio)chemicals, changing the reaction medium or the ionic or acidity level. These changes in the reaction medium affect the surface morphology and wettability in a way similar to physically stimulated systems.[19]

The general approach to obtaining a chemically responsive dynamic interface is to use pH-responsive polymers. The polymer layers can be generated in many different ways, including the layer-by-layer approach, the use of polymeric hydrogels, and electrostatic/covalent modification of the redox species in the polymer matrix. There are two main approaches to producing a pH-responsive interface. The first way is to add acidic or basic solutions to the reaction medium directly. Another way is by the in situ generation of acidity. Generally, researchers prefer the electrochemical oxidation of glucose in the presence of glucose oxidase, which produces gluconic acid and changes the local pH. In addition to changing the acidity in the reaction medium, the addition of different molecules can change the interfacial properties. In one study, researchers developed a cross-linked porous polymeric film using polyvinyl pyridine cross-linked with 1,4-diiodobutane on the surface of an indium tin oxide electrode. The designed polymeric interface exhibited reversibly opened and closed forms of the porous membrane in the absence and presence of cholesterol molecules, respectively. The presence of cholesterol results in the formation of hydrogen bonds on the polymeric chain and swelling of the polymer, which closes the pores. When the polymer is washed away, this hydrogen bonding is subverted and the system can return to the open formation. In this way, the charge-transfer resistance of the interface is controlled.[20,21]

In our recent study, we have developed a pH-responsive polymer material, poly(4-vinyl pyridine), which has the ability to respond to a change in the pH on the electrode surface for enzyme-based biocatalysis.[6] Basically, interfacial properties change due to protonation of the pyridine group in the polymer backbone under acidic conditions, which causes swelling of the polymer on the surface. This switchable swelling-shrinkage ability allows reversible diffusion of analyte through to the electrode surface. In our model system, we achieved control of the enzyme-based biomolecule under two different pH conditions (Fig. 1.10). In this system, when the pH of a medium is low (pH ≥ 6), the interface becomes impermeable to redox units and, as a result, electrode activity is turned off. However, when the polymer is charged as the result of a low pH (pH ≤ 5), the polymeric interface turns to be more permeable to ionic

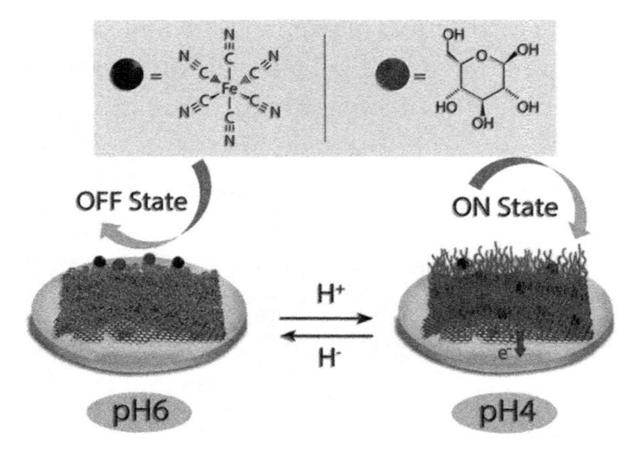

Figure 1.10 Representations of pH-encoded switchable graphene interfaces under two different pH conditions. Reprinted with permission from Ref. [6] Copyright © 2015 Royal Society of Chemistry.

species and allows their access to the conducting media, thus switching the electrode to the on state. In addition to changing the permeability and ionization ability of the polymer, morphology change also plays an important role in generating controllable interfaces.[6]

1.4 Programmable Bioelectronics

As mentioned, stimuli-responsive interfaces provide precise control over (bio)molecular interactions and signal output and also regulate inputs, including different molecules or different reaction conditions. However, current approaches are still not good enough to logically program all these interactions. Recent studies, especially in the area of biotechnology, show that there is an obvious need to develop programmable systems provide the ability to rearrange, control, and order their functions.

The programmable approach provides many opportunities to biotechnologists who specialize in artificial biological systems. It is well known that in nature, living reactions are naturally programmed, and attempts to construct artificial biological systems have generally resulted in vastly inferior performance. Because of this, in addition to advancement in the area of biotechnology to reach the target of artificial biological systems, the idea of biocomputing and/or programming is necessary.[1,4,21]

1.4.1 Enzyme-Based Logic Systems for Biocomputing

Initial works have used enzyme as a simple biological element for the processing of enzyme-based biocatalytic reactions expressed in the language of information processing.[22] Even though biocatalytic reactions are not really programmed, all interactions are defined in terms of programming language, which is crucial for initial steps. In this work, Boolean logic, which expresses the basic language of Boolean algebra using the physical implementation of logic functions, was used to define biocomputing systems. For example, various Boolean logic operations, such as AND, OR, XOR, NOR, NAND, INHIB, and XNOR, are employed to produce suitable designs. For all these Boolean logic operations, the only allowed values are defined as 0 and 1. These logic operations as a description of Boolean function are called "logic gates." In each logic gate, logic 0 is represented by any output signal below a certain threshold level and logic 1 is referred to be above this threshold level. Many different biocatalytic processes have been expressed in terms of Boolean logic operation, such as by using couple enzyme in solution in an optical system or by interfacing enzyme logic with signal-responsive materials as an electrochemical system.[23]

One of the most realistic biochemical computing studies has been developed using enzyme-based logic gates interfaced with signal-responsive materials. In this design, enzyme molecules are used as a biocatalytic input signal to process the information according to the Boolean functions AND/OR. Here, the AND-gate design works on the basis of a sequence of biocatalytic reactions starting with sucrose hydrolysis by invertase, which produces glucose. The glucose produced is oxidized by oxygen in the presence of glucose oxidase. The later reaction results in the formation of gluconic acid, which helps to reduce the local pH of the interface as the output of the gate. In this design, the absence and presence of enzymes are considered as input signals 0 and 1, respectively. In the AND gate, the biocatalytic chain reaction can only be active in the presence of both enzymes, glucose oxidase and invertase (1,1). The absence of one enzyme (0,1 or 1,0) or both of them (0,0) results in the absence of the gluconic acid formation, and the system does not produce a positive output signal. This system also requires an additional enzyme (urease) to reset the pH of the medium to the original value after reaction with urea. In the OR-gate design, the system works as two parallel reactions. The hydrolysis of ethyl butyrate and oxidation of glucose are catalyzed by esterase and glucose oxidase, and these reactions produce butyric acid and gluconic acid, respectively. The produced acid results in the reduction of the local pH and generates the signal. In this design, the input signals are two enzymes—glucose oxidase and esterase. The presence of both enzymes is considered

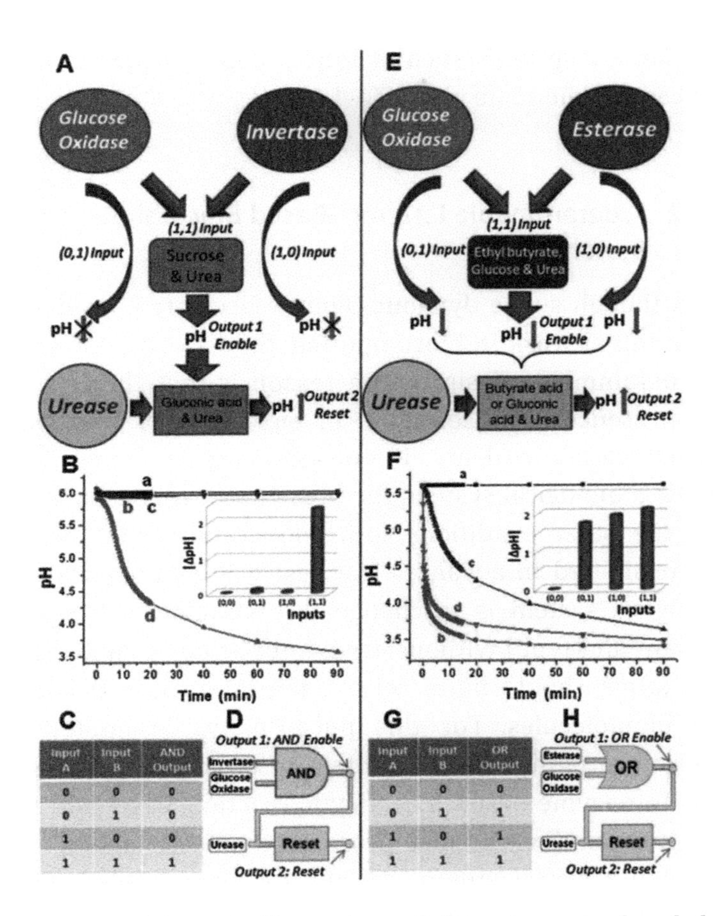

Figure 1.11 Schematic illustrations of the output signal and the truth table for AND-gate (a–d) and OR-gate designs (e–h), respectively. Reprinted with permission from Ref. [23] Copyright © 2015 Royal Society of Chemistry.

as (1,1). When either enzyme is absent, the system is considered as (0,1 or 1,0), and when both enzymes are absent, the system is considered as (0,0). In this study, the output signal is the pH of the medium, but it can be changed for different systems, such as optical or electrochemical, on the basis of the same idea (Fig. 1.11).[23]

1.4.2 Programmable Enzyme–Based Biocatalytic Systems

Even though single dynamic biointerfaces are considered as an initial work in biocomputing, they are still far from a programmable design.[1] The reason is that the idea of the presence or absence of an enzyme in a biological system is not realistic. If one attempts to design and mimic a biological system, the presence of every molecule and the other conditions (pH, temperature, etc.) should be considered in advance. The best way to program any biological system is to change the parameters (such as amount and type) without using interfering biomolecules. The target should be to control the interaction between the biomolecules. For this purpose, we attempted to design a programmable biointerface to control and regulate an enzyme-based biocatalytic reaction, segregate the enzymes, and eventually program their interactions by external physical stimuli. In our recent study, we have presented for the first time a design for a programmable interface with switchable and tunable biocatalytic performance that responds simultaneously to appropriate stimuli.[26]

In this model system, we have brought together temperature- and light-responsive polymers in two different combinations on an electrode surface. These designs allow us to control and regulate enzymatic reactions using simple Boolean logic operations.[26]

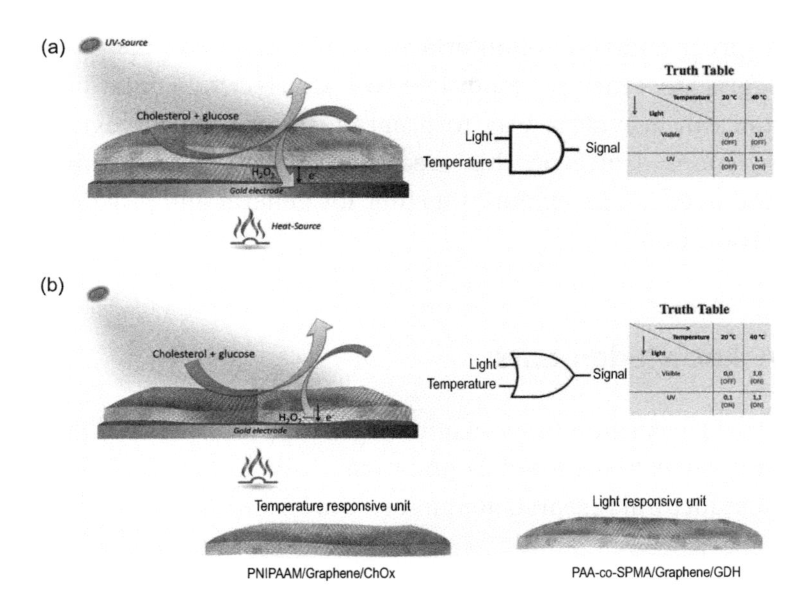

Figure 1.12 Logic gate designs for programmable bioelectrodes consist of temperature- and light-responsive polymers. Copyright © 2015 Royal Society of Chemistry.

1.5 Conclusion and Future Outlook

In conclusion, designing various switchable biointerfaces in order to understand the relationship between enzyme-based catalytic systems and changing environmental conditions on the basis of the electrochemical interface is useful and there is still a long way to go before critical issues can be solved, such as scaling the switchable interfaces up or down depending on the application and reversibility of the system. In addition to the generation of dynamic bioelectronic interfaces and related technology based on stimuli-responsive surfaces have evolved through a programmable biointerface design that can help to segregate reaction conditions on an electrode interface and can provide a way to program the rate of biological reactions. However, this area is also still in its infancy and needs to be explored further for general and practical applications.

Acknowledgments

Onur Parlak gratefully acknowledges support from the Knut and Alice Wallenberg Foundation (KAW 2014.0387) for postdoctoral research at Stanford University.

References

1. Parlak, O. (2016). Switchable bioelectronics, *Biosens. Bioelectron.*, **76**, pp. 251–265.
2. Turner, A. P. F. (2013). Biosensors: sense and sensibility, *Chem. Soc. Rev.*, **42**, pp. 3184–3196.
3. Willner, I. and Katz, E. (2006). *Bioelectronics: From Theory to Applications*, 1st ed. Wiley, USA.
4. Parlak, O. (2017). Interfacing graphene for electrochemical biosensing. In *Materials for Chemical Sensing*, eds. Paixao, T. R. L. C. and Reddy, S. M., Springer International, pp. 105–122.
5. Parlak, O., Beyazit S., Jafari, M. J., Tse Sum Bui, B., Haupt, K., Tiwari, A. and Turner, A. P. F. (2016). Light-triggered switchable graphene-polymer hybrid bioelectronics, *Adv. Mater. Interfaces*, **3**, pp. 1500353.
6. Parlak, O., Turner, A. P. F. and Tiwari, A. (2015). Ph-induced on/off-switchable graphene bioelectronics, *J. Mater. Chem. B*, **37**, pp. 7434–7439.

7. Parlak, O., Turner, A. P. F. and Tiwari, A. (2014). On/off-switchable zipper-like bioelectronics on graphene interface, *Adv. Mater.*, **3**, pp. 482–286.

8. Mendes, P. M. (2008). Stimuli-responsive surfaces for bio-applications, *Chem. Soc. Rev.*, **37**, pp. 2512–2529.

9. Jones, D. M., Brown, A. A. and Huck, W. T. S. (2002). Surface-initiated polymerizations in aqueous media: effect of initiator density, *Langmuir*, **18**, pp. 1265–1269.

10. Browne, W. R. and Feringa, B. L. (2009). Light switching of molecules on surface, *Annu. Rev. Phys. Chem.*, **60**, pp. 407–428.

11. Bocharova, V. and Katz, E. (2012). Switchable electrode interfaces controlled by physical, chemical and biological signals, *Chem. Rec.*, **12**, pp. 114–130.

12. Tunuguntla, R. H., Bangar, M. A., Kim, K., Stroeve, P., Grigoropoulos, C., Ajo-Franklin, C. M. and Noy, A. (2015). Bioelectronic light-gated transistors with biologically tunable performance, *Adv. Mater.*, **27**, pp. 831–836.

13. Rao, S. Y., Lu, S. F., Guo, Z. B., Li, Y., Chen, D. L. and Xiang, Y. (2014). A light-powered biocapacitor with nanochannel modulation, *Adv. Mater.*, **26**, pp. 5846–5850.

14. Huber, D. L., Manginell, R. P., Samara, M. A., Kim, B. I. and Bunker, B. C. (2003). Programmed adsorption and release of proteins in a microfluidic device, *Science*, **301**, pp. 352–354.

15. Pranzetti, A., Mieszkin, S., Iqbal, P., Rawson, F. J., Callow, M. E., Callow, J. A., Koelsch, P., Preece, J. A. and Mendes, P. M. (2013). An electrically reversible switchable surface to control and study early bacterial adhesion dynamics in real time, *Adv. Mater.*, **26**, pp. 2181–2185.

16. Katz, E., Minko, S., Halamek, J., MacVittie, K. and Yancey, K. (2013). Electrode interfaces switchable by physical and chemical signals for biosensing, biofuel, and biocomputing applications, *Anal. Bioanal. Chem.*, **405**, pp. 3659–3672.

17. Katz, E. (2010). Biofuel cells with switchable power output, *Electroanalysis*, **22**, pp. 744–756.

18. Katz, E., Sheeney-Haj-Ichia, L. and Willner, I. (2004). Electrical contacting of glucose oxidase in a redox-active rotaxane configuration, *Angew. Chem. Int. Ed.*, **25**, pp. 3292–3300.

19. Privman, M., Tam, T. K., Pita, M. and Katz, E. (2009). Switchable electrode controlled by enzyme logic network system: approaching physiologically regulated bioelectronics, *J. Am. Chem. Soc.*, **131**, pp. 1314–1321.

20. Contin, A., Plumeré, N. and Schuhmann, W. (2015). Controlling the charge of Ph-responsive redox hydrogel by means of redox-silent biocaalyric processes. A biocatalytic off/on switch, *Electrochem. Commun.*, **51**, pp. 50–53.

21. Tam, T. K., Zhou, J., Pita, M., Ornatska, M., Minko, S. and Katz E. (2008). Biochemically controlled bioelectrocatalytic interface, *J. Am. Chem. Soc.*, **130**, pp. 10888–10889.

22. Tam, T. K., Ornatska, M., Pita, M., Minko, S. and Katz, E. (2008). Polymer brush- modified electrode with switchable and tunable redox activity for bioelectronic applications, *J. Phys. Chem. C*, **112**, pp. 8438–8445.

23. Wang, L., Lian, W., Yao, H. and Liu, H. (2015). Multiple-stimuli responsive bioelectrocatalysis based on reduced graphene oxide/poly(N-isopropylacrylamide) composte films and its application in the fabrication of logic gates, *ACS Appl. Mater. Interfaces*, **7**, pp. 5168–5176.

24. Katz, E. and Privman, V. (2010). Enzyme-based logic systems for information processing, *Chem. Soc. Rev.*, **39**, pp. 1835–1857.

25. Parlak, O., Ashaduzzaman, M., Kollipara, S. B., Tiwari, A. and Turner, A. P. F. (2015). Switchable bioelectrocatalysis controlled by dual stimuli-responsive polymeric interface, *ACS Appl. Mater. Interfaces*, **43**, pp. 23837–23847.

26. Parlak, O., Beyazit S., Tse Sum Bui, B., Haupt, K., Turner, A. P. F. and Tiwari, A. (2016). Programmable bioelectronics in a stimuli-encoded 3D graphene interface, *Nanoscale*, **19**, pp. 9976–9981.

Chapter 2

Stimuli-Responsive Systems and Applications

Selim Beyazit
College of Engineering and Technology, American University of the Middle East, Kuwait
selimbeyazit85@gmail.com

2.1 Introduction

Stimuli-responsive (smart) polymers are unique materials that undergo dramatic physicochemical changes by responding to small differences in their environment.[1] The response can be solubility or charge transition, bond cleavage, conformation change, or dimerization. Stimuli can be categorized into two types, physical and chemical.[2] Chemical stimuli, such as pH, ionic agents, and chemical agents, will change the interaction between macromolecules or macromolecules and solvent. Physical stimuli, like temperature, magnetic field, and light, can cause more

Switchable Bioelectronics
Edited by Onur Parlak

ISBN 978-981-4800-89-1 (Hardcover), 978-1-003-05600-3 (eBook)
www.jennystanford.com

complex effects at the molecular level, resulting in drastic change at the macroscopic level. Recently, materials that respond to enzymes, antigens, and ligands were reported and categorized as biochemical-responsive materials.[3]

Responsive polymers can be used in many applications, such as drug delivery,[4] biotechnology,[5] and chromatography,[6] in different forms; Stimuli-responsive polymers can respond to a single stimulus or more than one stimulus, such as the combination of heat and pH.[7] Stimuli-responsive polymers can be used in different physical forms; that is, hydrogels (cross-linked polymers), micelles (linear polymers), and modified interfaces. Hydrogels are 3D polymeric networks that can absorb water. Some parts of the hydrogels are solvated by water, while the rest are bound to each other by chemical or physical bonds. Micelles are constituted of an agglomeration of amphiphilic polymers. They contain hydrophobic and hydrophilic segments. The hydrophobic segments aggregate and make nano-/micromicelles in water, while the hydrophilic segments stabilize these aggregates in hydrophilic solvents. The same approach can be used in hydrophobic solvents that solubilize the hydrophobic segments while the hydrophilic parts aggregate. In the case of interfaces, polymers can be grafted onto a solid surface (polymeric surface, metal surface, glass, etc.) by various methods.

In this chapter, the fundamentals and basic applications of stimuli-responsive polymers will be reviewed.

2.2 Thermoresponsive Materials

Heat is widely used as a stimulus due to the eases of its application either in vitro or in vivo. The essential property

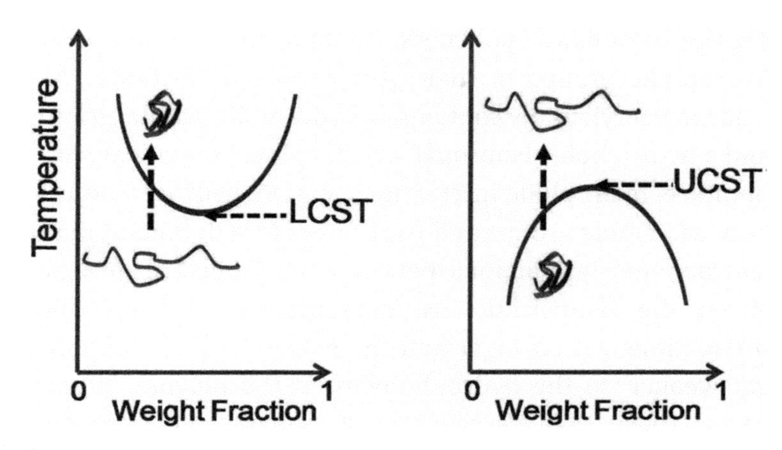

Figure 2.1 Phase transition associated with a lower critical solution temperature (LCST) and an upper critical solution temperature (UCST) behavior. Reprinted from Ref. [8].

of thermoresponsive polymers is the existence of a critical solution temperature. The critical temperature refers to the temperature at which phase change of the polymer and solvent occurs. If the polymer is soluble in water and it turns hydrophobic above a certain temperature, this polymer has a lower critical solution temperature (LCST). If the polymer is not soluble in the solvent and it turns soluble above a certain temperature, it means that this polymer has a higher critical solution temperature, also known as an upper critical solution temperature (UCST) (Fig. 2.1).

Although there are two types of critical solution temperatures, materials that have LCSTs are used much more frequently. Commonly, polymers are more soluble at elevated temperatures. However, polymers that have LCSTs turn hydrophobic above their LCSTs. The reason behind this phenomenon is the competition between hydrophobic and hydrophilic interactions on the polymers.

Thermoresponsive polymers contain hydrophobic and hydrophilic groups in their structures. For instance, *N*-isopropylacrylamide contains a hydrophilic amide group and a hydrophobic isopropyl group. Below the critical temperature, hydrophilic interactions, mainly hydrogen bonds between amide groups and water, are more dominant than hydrophobic interactions between the isopropyl groups. When the temperature is increased, the hydrophobic interactions get stronger and the hydrophilic interactions get weaker, so the hydrophobicity of the polymer dominates. These changes cause a total phase change of the polymer and make it nonsoluble above the LCST. Moreover, the LCST of polymers can be tuned by adding more hydrophobic or hydrophilic functionalities to the polymeric segments. For instance, the LCST of polyacrylamide-*co*-poly(*N*-isopropylacrylamide) (PAAm-*co*-PNIPAM) is higher than that of the homopolymer of PNIPAM. If more hydrophobic groups are introduced, the LCST of PAAm-*co*-PNIPAM will be lower than that of PNIPAM alone (Fig. 2.2).

In addition to poly(*N*-substituted acrylamides), which are well-known thermoresponsive polymers, other types of thermoresponsive polymers—such as poly(ethylene oxide)-poly(propylene oxide)-poly(ethylene oxide) (PEO-PPO-PEO) triblock copolymers—exhibit temperature-responsive micellization and gelation. They have a sol-gel phase transition at around 50°C.[9] Moreover, some biopolymers, like agarose, gelatin, and gellan benzyl ester, show thermoresponsivity. Gellan is a polysaccharide that forms helix conformations via hydrogen bonding in aqueous systems. When the temperature is increased to a value above the LCST, hydrophobic interactions on the chains

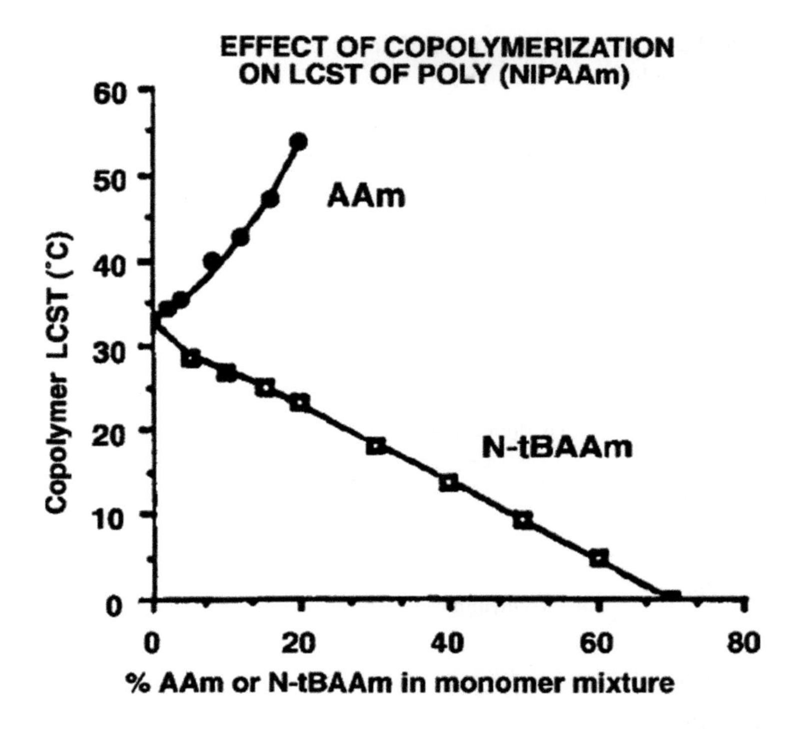

Figure 2.2 Effect of copolymerization of poly(*N*-isopropylacrylamide) with acrylamide (more hydrophilic comonomer) and *N*-*tert*-butylacrylamide [3].

are dominant and the gellan aggregates in water. Chemical structures of some common thermoresponsive polymers are shown in Table 2.1.

Micelles can be obtained by using amphiphilic block copolymers, which means that they contain both hydrophilic and hydrophobic segments in one polymer chain. If at least one of the segments is thermoresponsive, the micellation process can be controlled by using thermal changes. Armes and Liu reported the synthesis of poly(propylene oxide)-2-(dimethylamino)-(ethyl

Table 2.1 Chemical structures of some common LCST thermoresponsive polymers

Structure	Name	Abbreviation
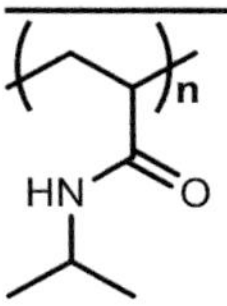	Poly(*N*-isopropylacrylamide)	PNIPAM
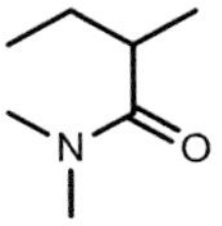	Poly(*N*-dimethylacrylamide)	PDMAm
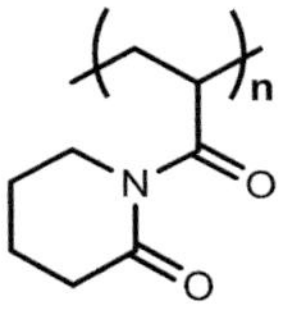	Poly(*N*-vinylpiperidone)	PVPip
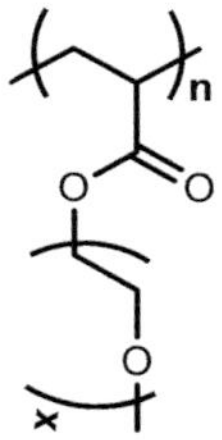	Poly[oligo(ethyleneglycol)-methacrylate]	POEGMA

methacrylate)-oligo(ethylene glycol)methacrylate (PPO-DMAEM-OEGMA) ABC triblock copolymers via atom transfer radical polymerization (ATRP), which undergoes reversible micellization.[10] The addition of 1,2-bis(2-iodoethoxy)ethane, a bifunctional quaternizing agent, results in cross-linking of the DMAEM units at 40°C and pH 8.5 (Fig. 2.3). The authors prepared different types of nanoparticles (i.e., tunable micelle structures with variable

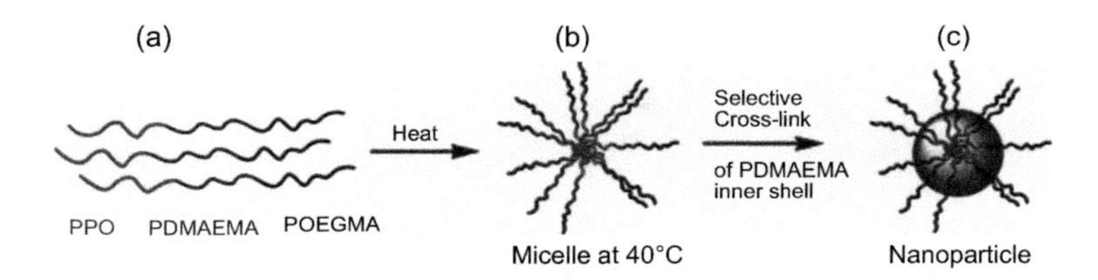

Figure 2.3 Schematic drawing of the aqueous solution of a molecularly dissolved triblock copolymer at 5°C (a); and the formation of micelles at 40°C (b); selective cross-linking of inner-shell permanent nanoparticle (c). Adapted from Ref. [11].

degrees of swelling) from the same triblock copolymer by varying synthesis temperature, copolymer concentration, and cross-linking degree.

Thermoresponsive hydrogels are cross-linked networks that can shrink or collapse above the LCST. The change of the network morphology results in the loss of water and any cosolutes, such as therapeutic agents, from the hydrogel. If the drug is loaded onto a hydrogel in its swollen state (temperature lower than the LCST of the polymer), the increase of the temperature to a value above the LCST results initially in fast drug release due to a sharp and drastic change in the structure of the network, followed by a slow release of the therapeutic agent. Kaneko et al. reported the release of the sodium salt of salicylic acid as such.[12] Applications of thermoresponsive polymers for drug delivery have been extensively studied and reviewed several times (Fig. 2.4).[1,2,11]

The interaction between macromolecules and solvent can be manipulated by other molecules, such as salts and surfactants. These additives can cause a dramatic LCST or a total disappearance of the critical surface temperature. For instance, the addition of an ionic surfactant, such

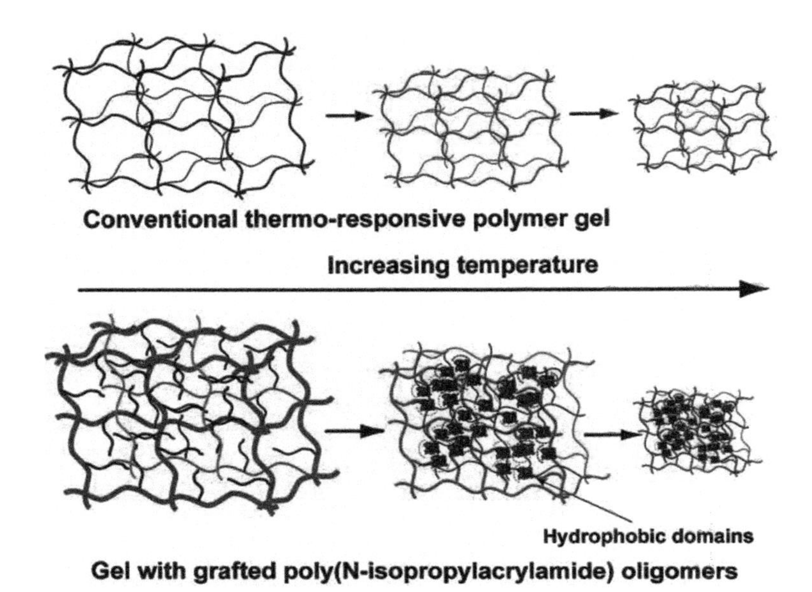

Figure 2.4 Architectures of thermoresponsive gels. Adapted from Ref. [11].

as sodium dodecyl sulfate, causes an increase in the hydrodynamic radius of PNIPAM whereas the radius of PVC decreases.[13] Although PNIPAM is the most abundant thermoresponsive polymer, there are some concerns about its biocompatibility, phase transition delays, and effect of end groups on thermal behavior.[14] Thus, other thermoresponsive polymers are still under investigation. For instance, poly(*N*,*N*-diethylacrylamide) (PDEAM) is a thermoresponsive polymer that has a thermal transition around 33°C near PNIPAM. Bradley et al. showed the application of PDEAM in nanomechanical cantilever sensors.[15]

Thermoresponsive polymers are used in gene delivery applications. Gene therapy aims to treat genetic diseases by correcting deflecting genes that cause diseases with therapeutic genes (DNAs). DNAs are negatively charged, hydrophilic molecules that are supposed to reach the nucleus of the cell, which is negatively charged and hydrophobic. PNIPAM, which is grafted to a cationic polymer such as polyethyleneimine or chitosan, can interact electrostatically with therapeutic DNA and allow effective transfection.[16,17]

2.3 Systems That Are pH Responsive

Polymers that are pH responsive respond to changes in the pH of the environment. All pH-responsive polymers contain pendant proton absorbing or releasing groups in their polymer chains. These groups can be acidic (sulfonic or carboxylic acids) and can release their protons and turn anionic at high pH values and neutral at low pH values. The other way is to use basic moieties, like ammonium salts, that can be protonated at low pH values and deprotonated at high pH values. In addition to the total charge, the morphology of the polymer also changes due to charge repulsion in the network. For instance, a carboxylic acid–based pH-responsive polymer is in its collapsed state under a certain pH value and turns anionic and swells due to charge repulsion when the pH is elevated. In addition, pH can be used to modulate the solubility of the polymer (Fig. 2.5).

Polymers that are pH responsive have been frequently used for the controlled release of orally administered drugs. The pH of the stomach is highly acidic (<3), the pH of the intestine is neutral, and the pH of the colon is basic. The use of polycationic polymers prevents the

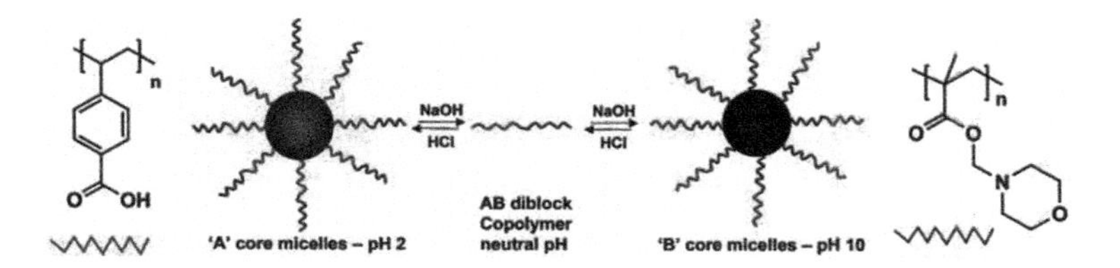

Figure 2.5 Control of micellar states dependent on pH. Adapted from Ref. [11].

release of the therapeutic agent in the mouth due to a neutral pH. Polycationic polymers are neutral at this stage and the release of the drug is minimum. In the stomach, the polymer turns to a cationic state, which causes swelling of the polymer, resulting in fast release of the therapeutic agent in the stomach. When caffeine was loaded into hydrogels made of copolymers of methyl methacrylate and DMAEM, it was not released at a neutral pH but released at zero order at a pH of 3–5, where DMAEM became ionized.[18] A polyanionic hydrogel with an azoaromatic cross-linker can be used for colon-targeting drug delivery. Drug release from polyanionic hydrogels in the stomach is minimum. Although the polymer is swollen in intestines, its degradation occurs only in the colon due to the azoreductase produced by the microbial flora of the colon (Fig. 2.6).[19]

Polymers that are pH responsive have also been used in sensing applications. In general, they are combined by enzymes that can change the pH of the environment of the polymers when they interact with their substrate. A well-known example of pH-responsive biosensors is glucose oxidase–loaded pH-responsive polymers for the detection of glucose. Glucose oxidase converts glucose to gluconic acid, and the pH of the environment starts decreasing,

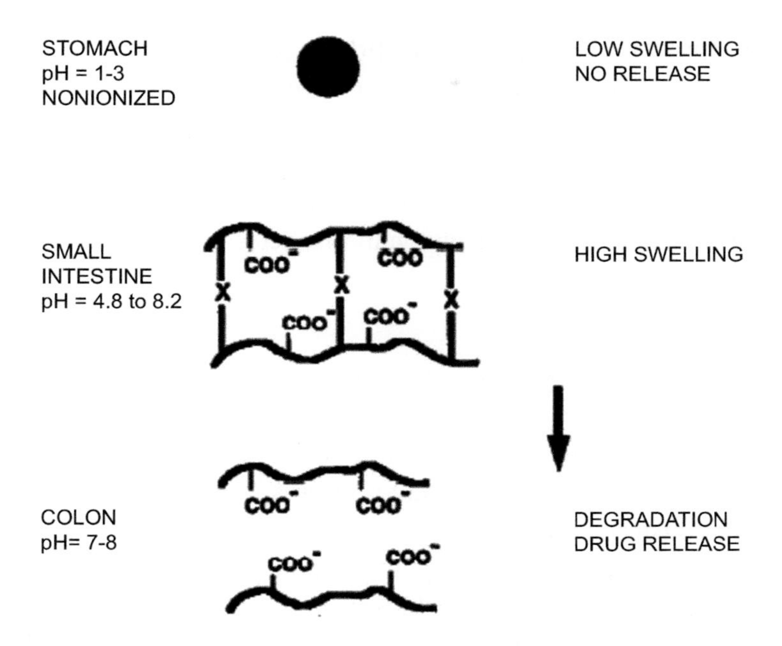

Figure 2.6 Schematic illustration of oral colon-specific drug delivery using biodegradable and pH-sensitive hydrogels. Adapted from Ref. [19].

resulting in the conversion of the neutral polymer into a cationic electrolyte and simultaneous swelling of the polymers. Mechanisms that are pH responsive can be used at the tissue level. Human tumors exhibit acidic pH levels that range from 5.7 to 7.8.[20] Accumulation of lactic acid in rapidly growing tumor cells is the main reason of the acidity of tumor microenvoriments.[21] Griset et al. used acrylate-based nanoparticles that bear hydroxy group.[22] The hydroxy groups were masked by a pH-sensitive protecting group (2,4,6-trimethoxybenzaldehyde), which results in hydrophobic polymeric nanoparticles. These nanoparticles are stable in a neutral pH, but protecting groups start cleaving when the pH is slightly acidic (pH $\approx$5), which results in the transition of the nanoparticles from hydrophobic to hydrophilic and subsequent drug release. This system shows an efficient inhibition of the rapid growth of Lewis lung carcinoma tumors in C57Bl/6 mice compared to nonresponsive nanoparticles or paclitaxel in solution.

2.4 Photoresponsive Systems

Photoresponsive polymers contain special functionalities that respond to light. Typically, photoresponsive functionalities cause chemical or physical change of macromolecules. Possible applications of photoresponsive polymers are reversible optical storage biosensors, drug delivery, and bioactivity switching of activity of proteins.[23] The most important advantage of light-responsive polymers is the relatively straightforward application of light to induce a responsive behavior.

Photoresponsive molecules can be divided into three categories; (i) photoisomerization, (ii) photodimerization and (iii) photocleavage. A few of these mechanisms are reversible and can be repeated several times, in which a reverse reaction occurs either under irradiation at different wavelength or by using different stimuli, like temperature.

2.4.1 Photoisomerization

Commonly, photoisomerization is a reversible and repeatable process. These features make it very attractive for many applications. Using molecules such as azobenzene and spiropyran (SP) moieties exemplifies this process.

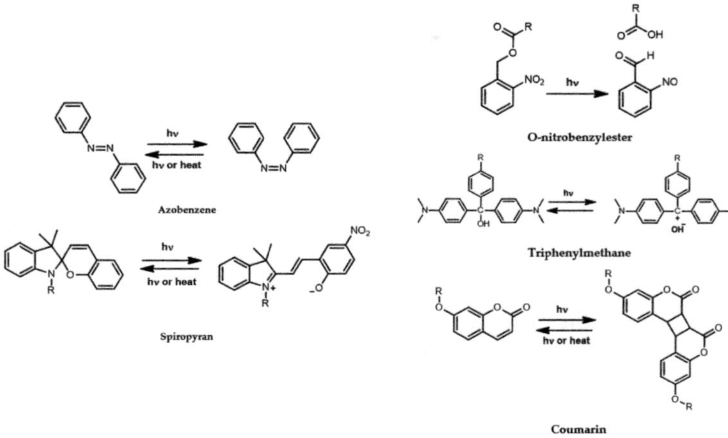

Figure 2.7 Representative photoactive groups.

2.4.1.1 Azobenzene-based systems

Azobenzenes can switch from *trans* to *cis* by ultraviolet (UV) irradiation, and the *cis* form turns back to its thermodynamically stable *trans* conformation by visible light irradiation or heat (Fig. 2.7 top). The conformation change is accompanied by the fast and complete change of the electronic structure, geometric shape, and polarity. For instance, while the more stable *trans*-azobenzene has no dipole moment, the *cis* form is quite polar, having a dipole moment of 3 Debye. Azobenzene moieties can be used to control the hydrophobicity/hydrophilicity of the polymer chain. In addition to polarity, a change in the conformation of the azobenzenes can induce steric hindrance, which results in a change in the polymer morphology.

2.4.1.1.1 *Supramolecular cross-linking*

Kros and coworkers reported the synthesis of a supramolecular gel for protein release.[24] First, maleimide

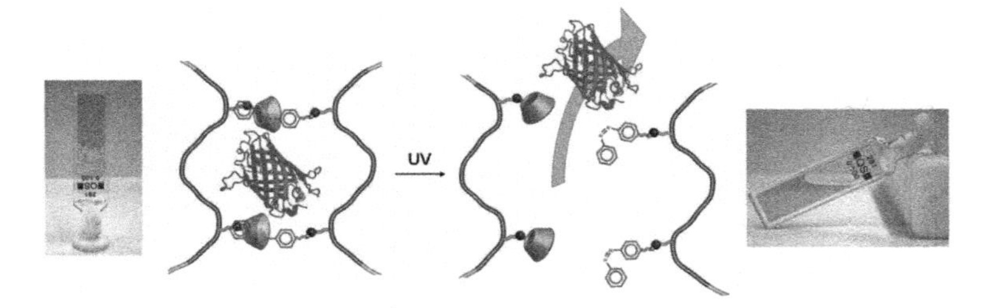

Figure 2.8 Schematic representation of a photoresponsive protein released from the hydrogel composed of *trans*-azobenzene-modified dextran and cyclodextrin-modified dextran. Adapted from Ref. [24].

moieties were grafted onto dextran, a natural polymer. Then, maleimide-grafted dextrans were functionalized with either cyclodextrin or azobenzene moieties separately. The two resulting linear polymers were used as the building blocks of supramolecularly cross-linked hydrogels for the light-controlled release of proteins. As stated before, *trans* azobenzene is hydrophobic and it can be trapped in the interior part of cyclodextrins. Upon irradiation with UV light, conformation change from *trans* to *cis* causes deformation of the physically cross-linked gels and the simultaneous release of proteins (Fig. 2.8).

Gupta and coworkers used a similar method to obtain nanogels for drug delivery.[25] Instead of using an azobenzene-cyclodextrin complex as the supramolecular cross-linker they used hydrophobic interaction between azobenzene moieties. Aspirin-loaded nanogels were irradiated with UV light, and hydrophobic interactions were weakened by conformation change. The resulting loose structure allowed the release of aspirin. In addition, they showed the effect of pH on the release profile of the nanogels. The drug was released almost 2 times faster at a pH of 9 than at a pH of 4. Cytotoxicity studies revealed that the nanogels were toxic to cells above 1 mg/mL.

2.4.1.1.2 *Self-assembled systems*

Azobenzene groups can also be used in block copolymers to regulate the hydrophilicity of one segment. A block copolymer that contains a hydrophilic segment and a hydrophobic azobenzene-bearing segment can be used to obtain self-assembly micelles. The hydrophobic azobenzene segment will aggregate due to hydrophobic interaction, and the hydrophilic segments will stabilize the micelles. Recently, Guo et al. used a similar method to obtain triple responsive polymers.[26] They synthesized pH-, temperature-, and photoresponsive polymers. The hydrophilic segment of the polymer contains poly(2-(dimethylamino)-(ethyl methacrylate)-*co*-6-*O*-methacryloyl-D-galactopyranoside), or P(DMAEM-*co*-MAGP), for thermal and pH responsiveness. The hydrophobic segment contains hydrophobic poly[4-(4-methoxyphenylazo) phenoxy methacrylate] (PMAZO). The size of the micelles was regulated by changing the pH of the environment and the temperature. The authors showed the effect of different pH values on the temperature responsivity. Till pH 7, there is no thermoresponsivity because DMAEM is totally protonated and highly soluble. At pH 8, the cloud point is around 50°C. Increasing the pH causes the lowering of the LCST due to a more hydrophobic nature of the polymer at an elevated pH. The authors also controlled the photoresponsivity of the micelles. The size of the micelles was followed by transmission electron microscopy (TEM) and dynamic light scattering. Without irradiation, the size of the micelles was around 100 nm. After UV irradiation, the azobenzene

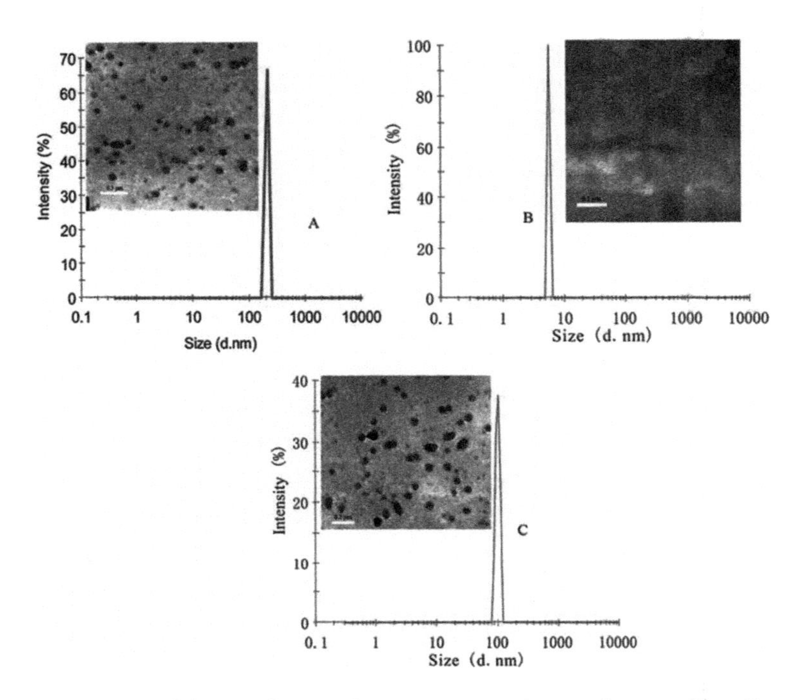

Figure 2.9 (a) DLS plots and TEM images of P(DMAEM-*co*-MAGP)-*b*-PMAZO micelle solution before UV irradiation (A), after UV irradiation (B), and after exposure to visible light (C), at pH 3. Adapted from Ref. [26].

moieties changed to a more hydrophilic conformation and resulted in the deformation of the micelles. After irradiation by visible light, the azobenzene groups turned back to their hydrophobic *trans* conformation and micelles were reobtained (Fig. 2.9). These studies were done at a pH of 3, where P(DMAEM-*co*-MAGP) is hydrophobic. The authors did not show any data at higher pH values, but at pH 11, completely reversed behaviors can be obtained. Since the first segment is more hydrophobic at pH 11, no micellization is supposed to occur without UV light irradiation and micelles can be obtained just after UV light irradiation.

2.4.1.1.3 *Nanoimpellers*

Apart from applications derived from their physicochemical properties, azobenzenes have been exploited for their motional behaviors. With conformational change, they can be used like a mechanical stirrer (nanoimpeller). Zink and coworkers reported the synthesis of light-activated nanoimpeller-controlled drug release from mesoporous silica nanoparticles.[27] Light of 457 nm was used to irradiate and activate the system where *cis* and *trans* conformations of azobenzenes are absorbed. In addition, the behavior of this system was investigated in cancer cells. A second contribution from the same group showed the activation of the same process at 760 nm by using a two-photon upconversion fluorophore.[28] The fluorophore possesses a high two-photon absorption cross section that emits close to the absorption of azobenzene groups. The *trans*-to-*cis* isomerization of azobenzenes occurs by the fluorescence resonance energy transfer (FRET) process (Fig. 2.10). Fluorophore groups and azobenzene groups were introduced into mesoporous silica particles by their polymerizable silane units. Drug release profile of the nanoparticles were demonstrated by using by one-photon excitation process (λ_{ex} = 365 nm) and two-photon excitation process (λ_{ex} = 760 nm).

A similar approach was reported by Liu et al., but in this case upconverting nanoparticles (UCPs) were used instead of two-photon excitation possessing fluorophore. UCPs have emission bands around 350 nm, which activates the change from *trans* to *cis* conformation, and around 450 nm,

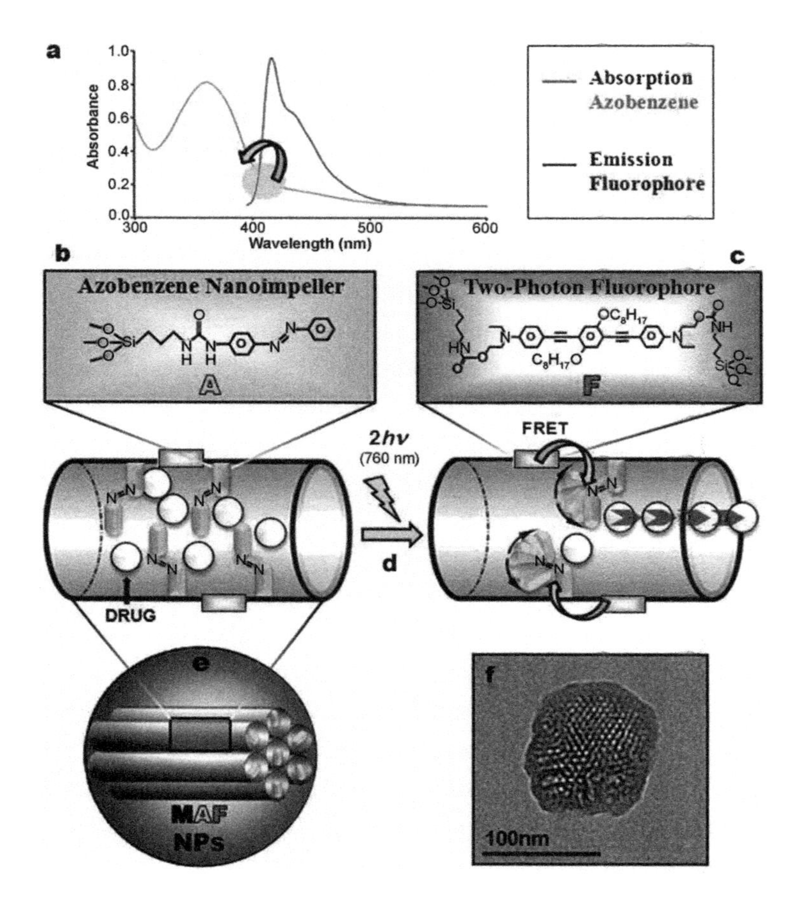

Figure 2.10 Mesoporous silica nanoparticles combining (b) azobenzene moieties and (c) a two-photon fluorophore. The design of the so-called MAF nanoimpellers allows a two-photon (760 nm)-activated release of drug molecules by (a) FRET and (d) photoisomerization of azobenzene. (e) Nanoimpellers and (f) their transmission electron microscopy image. Adapted from Ref. [28].

which activates the change from *cis* to *trans* conformation. They synthesized a mesoporous silica shell that bears azobenzene functionalities around UCPs. Irradiation with

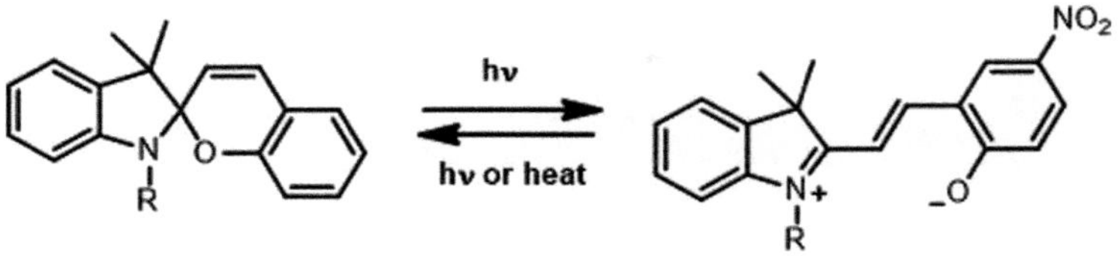

Figure 2.11 Photoisomerization of spiropyran.

light of 980 nm activated the nanoimpellers and resulted in the release of the loaded cargo. The authors showed the fate of the nanoparticles in the cells by confocal laser scanning microscopy and cell viabilities. There was no cell death with control experiments and the number of dead cells increased with increasing infrared irradiation time. This result proves effective and controlled release of doxorubicin from nanoparticles.

2.4.1.2 Spiropyran-based systems

SP derivatives are widely used photochromic compounds found in many applications, such as supramolecular assembly, photoswitchable nanoparticles, photoswitched binding of DNA or RNA, and photoresponsive polymers (Fig. 2.11).

2.4.1.2.1 *Supramolecular assembly*

SP derivatives change their structures from SP to merocyanine (MC) upon UV light irradiation. The reversible isomerization between the uncharged and nonplanar SP form and charged, planar MC derivative can be regulated by UV light (SP to MC) or visible light (MC to SP). Qiu et al. used the planar structure of MC for the synthesis of photoswitchable hydrogels by using p-p interaction between MCs.[29] First, they introduced a D-Ala-D-Ala dipeptide to a SP compound. A solution of this compound at pH 3 was irradiated with

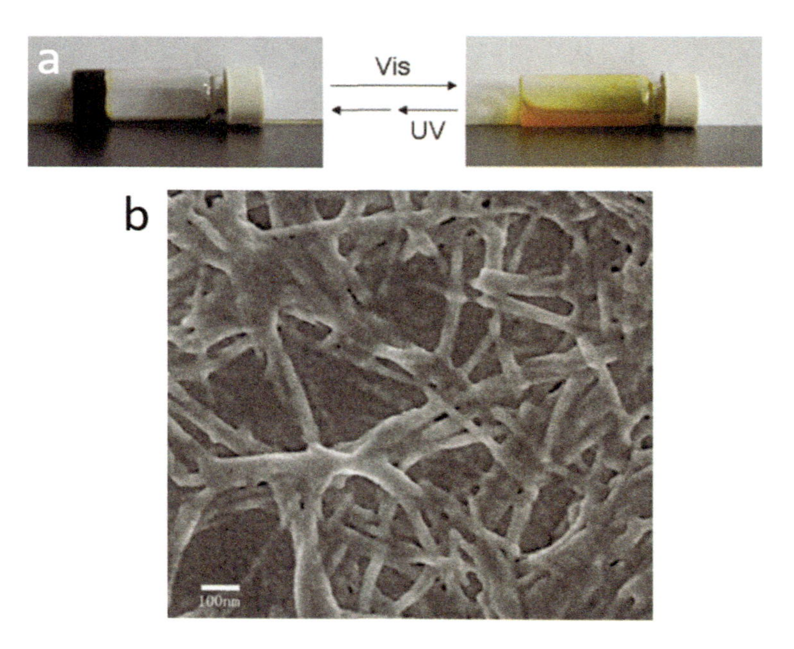

Figure 2.12 (a) Photoresponsivity and (b) SEM image of the hydrogel formed by merocyanine. Adapted from Ref. [29].

UV light, and the solution turned to a gel. The reversibility of the system was controlled by irradiating visible light, which resulted in a yellow slurry (Fig. 2.12a). In addition to that, increasing the pH prevented any sol-gel behavior. The authors didn't explain clearly, but this behavior happened probably due to the more hydrophilic nature of MC. A higher pH causes conversion of phenol moieties to phenoxide groups on the MC, providing no gelation, but at pH 3, MCs contain phenol groups that are less soluble than phenoxide salt. Moreover, the authors showed ligand sensitivity of the system by using vancomycin, for which the D-Ala-D-Ala dipeptide has a strong affinity. The gel turned into a red solution upon addition of the equivalent amount of vancomycin hydrochloride onto the surface of the gel. They explained that this behavior was due to a more hydrophilic nature of the ligand, which destroys the hydrophilic-hydrophobic balance of the system.

2.4.1.2.2 *Self-assembly*

SP derivatives can be used to control association/dissociation of self-assembled micelles due to hydrophobic/hydrophilic change upon UV irradiation since the MC isomer is hydrophilic and SP is hydrophobic. Matyjaszewski and coworkers reported the synthesis of photoresponsive amphiphilic block copolymers by ATRP.[30] A PEO macroinitiator was used as the hydrophilic segment, and the subsequent polymerization of SP methacrylate by ATRP resulted in a photoresponsive amphiphilic block copolymer. They showed the dissociation of the micelles upon UV irradiation and the reassociation of micelles after visible light irradiation. The obtained micelles were used for the photoregulated release and capture of a coumarin-based dye. A similar study was reported by Zhou et al., who synthesized polyspiropyran methacrylate-*b*-polyacrylic acid by copper(0)-mediated living radical polymerization.[31] Since copper-mediated polymerizations are sensitive to acidic moieties, the authors used *t*-butyl acrylate to synthesize block copolymers and hydrolyzed *t*-butyl acrylate to methacrylic acid. The obtained polymers showed photoresponsive and pH-responsive properties that result in a "schizophrenic" behavior. Without any treatment, neither photo nor pH block copolymers assembled themselves into micelles that contain hydrophobic SP units and hydrophilic acrylic acid units. UV light treatment induced a conformation change from SP to MC that caused dissociation of the micelles since the polymer turned totally hydrophilic. Visible light irradiation caused conformation change from MC to SP that resulted in micellization again due to the hydrophobicity of SP units. A decrease in the pH and UV light irradiation resulted in protonation of acrylic acid, which is more hydrophobic, and conformation change to a hydrophilic MC isomer, causing micellization again.

The schizophrenic behavior can be obtained by different groups responsive to a combination of different stimuli. For instance, the combination of light-responsive SP methacrylate and thermoresponsive di(ethylene glycol) methyl ether methacrylate (DEGMMA) was reported by Jin et al.[32] (Fig. 2.13). In this case, the self-assembly of the block copolymers were controlled by light and heat and similar behaviors were observed as mentioned above.

Feng et al. reported the synthesis of photoresponsive, random, amphiphilic copolymers by ring-opening metathe-

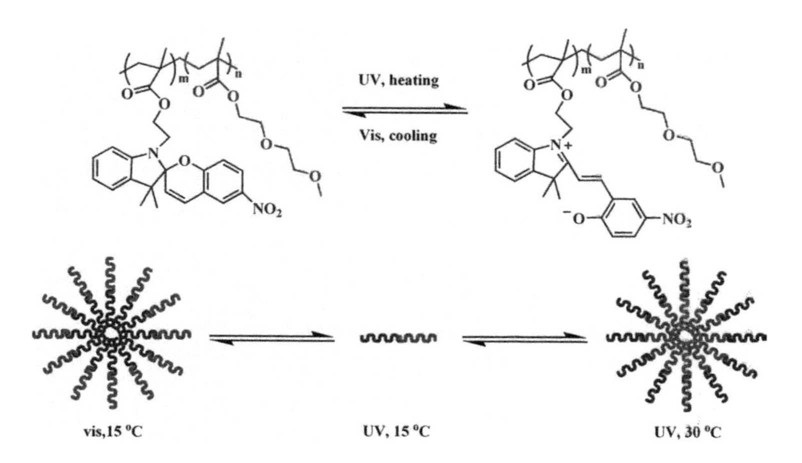

Figure 2.13 Schematic representation of "schizophrenic" behavior of PSPMA-*b*-PDEGMMA. Adapted from Ref. [32].

sis polymerization.[33] Norbornane-bearing hydrophilic, hydrophobic, and photoresponsive monomers were polymerized by using Grubbs' third-generation initiator. The self-assembly behavior and the drug loading/release profile of the obtained random copolymer were shown. Upon UV light radiation, the obtained micelles were disrupted and subsequent visible light irradiation led to the reformation of the micelles.

2.4.2 Photodimerization

Photodimerization refers to reversible coupling of identical molecules by light irradiation. The major advantage of this system is that it can change the cross-linking density and by doing so it can reversibly change the physical properties of the polymers. Well-known examples of photodimerizable moieties are coumarins,[34] anthracene (AN),[35,36] and cinnamoyl[37] derivatives.

2.4.2.1 Anthracene derivatives

AN chromophore groups can be dimerized on treatment with light or heat through a [4+4] cycloaddition reaction (Fig. 2.14). Wells et al. reported the effect of the photodimerization of AN on the physicochemical properties and drug release behavior of hydrophilic polymers.[35] They introduced pendant AN units onto NH_2-PEO chains. The resulting NH_2-PEO-AN chains were grafted onto alginate

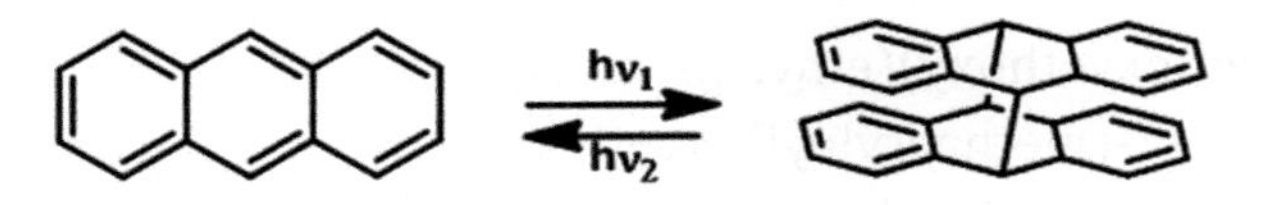

Figure 2.14 Dimerization of anthracene.

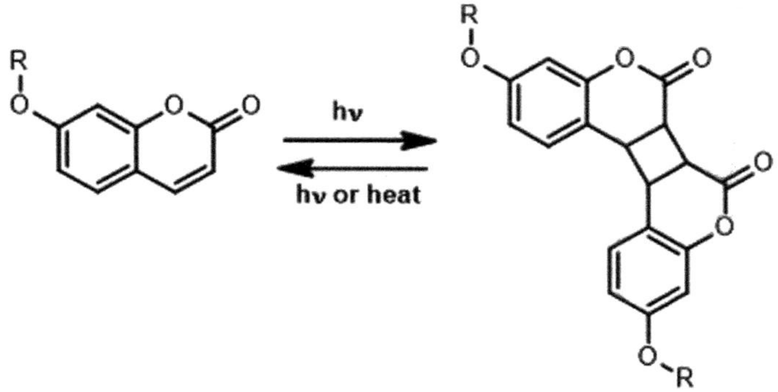

Figure 2.15 Reversible dimerization of coumarin derivatives.

or the hyaluronate backbone. UV light irradiation caused photodimerization of ANs, which changed the release profile of the loaded cargo.

2.4.2.2 Coumarin derivatives

Coumarins are members of the benzopyrone family derived from cinnamic acids. Dimerization of coumarins requires light above 300 nm, as below 300 nm, dimerized coumarins are cleaved (Fig. 2.15).

Coumarins are used in a manner similar to the way ANs are used. The photodimerization of coumarin groups changes the physicochemical properties of the polymers, such as their swelling degree, cross-linking density, and glass transition temperature. He et al. used coumarins to obtain multiresponsive nanoparticles.[38] They used halide-functionalized PEO for random copolymerization of 2-(2-methoxyethoxy)(ethyl methacrylate) (MEOMA) and 4-methyl-[7-(methacryloyl)oxyethyloxy]coumarin (coumarin methacrylate [CMA]) by ATRP. The resulting polymer, PEO-*b*-PMEOMA-*co*-CMA, self-assembled to micelles above its

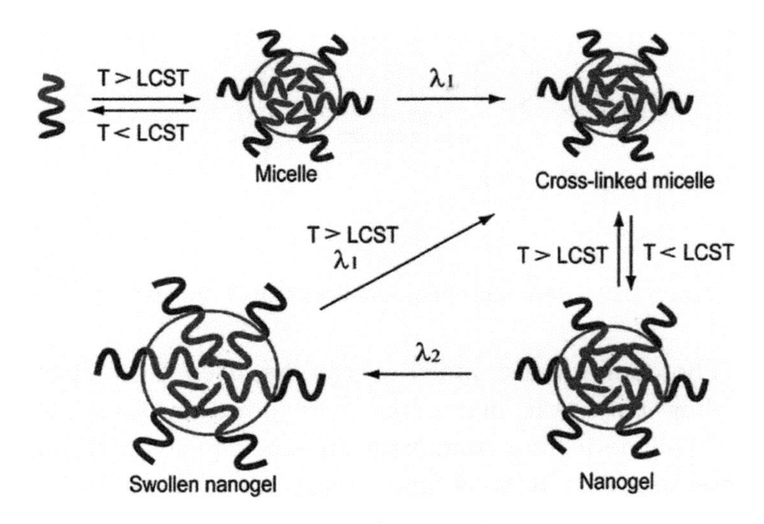

Figure 2.16 Schematic illustration of the preparation and photo-controlled volume change of PEO-*b*-PMEOMA-*co*-CMA. Adapted from Ref. [38].

LCST. Irradiation with light of 310 nm at above the critical temperature of the polymer resulted in cross-linked micelles, which still showed thermoresponsive properties. Above the LCST, the cross-linked micelles shrink and below this temperature, they expand. In addition, irradiation with light of 260 nm caused dissociation of the majority of the coumarin dimers, resulting in a looser core structure (Fig. 2.16).

2.4.2.3 Cinnamoyl derivatives

Cinnamoyl derivatives undergo dimerization under heat or light. Above 300 nm, cinnamoyl groups photodimerize, and subsequent irradiation with light of 260 nm results in the dissociation of the dimers (Fig. 2.17).

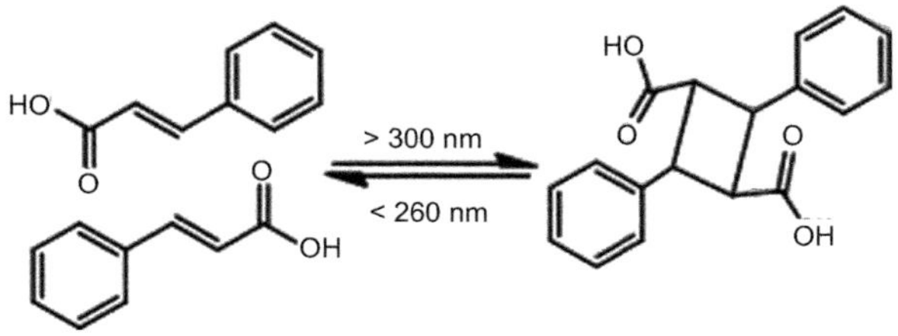

Figure 2.17 Reversible photodimerization of cinnamic acid.

Photodimerizable moieties can be used to prepare healing polymeric materials, as reported by Chung et al.[39] Polymeric films containing cinnamoyl pendant groups were prepared initially, and cracks were made on the film. Irradiation with a 260 nm light source resulted in the dissociation of cinnamoyl dimers and subsequent irradiation with light >280 nm generated re-cross-linking of the cinnamoyls and healing of the cracks. The efficiency of the de-cross-linking/re-cross-linking procedure was followed by Fourier-transform infrared spectroscopy, and the healing ability of the method was controlled by tensure strength of the film.

2.4.3 Photocleavage

Photocleavable or photolabile-based polymers are very useful materials in biomedical applications. The most well-known examples are *o*-nitrobenzyl (ONB) alcohol derivatives. They can be photocleaved under appropriate light irradiation, resulting in few side products (Fig. 2.18).[40,41] Moreover, they are very stable in acidic and basic conditions. In addition to ONB compounds, *o*-nitrophenyl ethyl alcohol derivatives also show the same behavior. Kim et al. compared the photoactivity of ONB and *o*-nitrophenyl

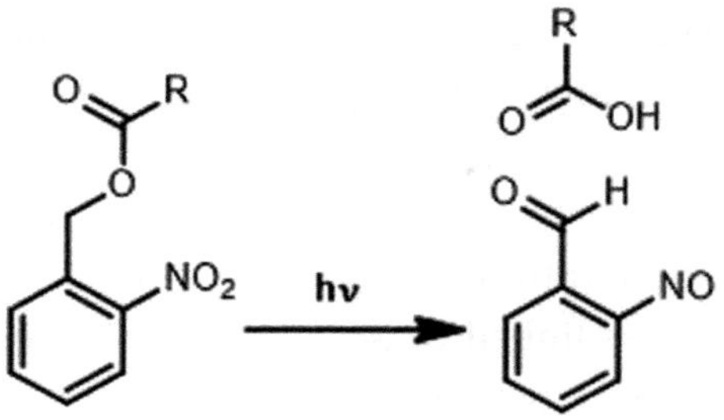

Figure 2.18 Photocleavage of *ortho*-nitrobenzyl derivatives.

ethyl alcohol derivatives and reported that *o*-nitrophenyl ethyl alcohol derivatives were cleaved significantly faster than ONB alcohol derivatives.[42] However, ONB derivatives are more widely applied.

ONB derivatives have been intensely used for controlled drug release studies. They can be used for either disruption of the carrier morphology, which enables release of the drug, or direct photocleavage of the covalently bound drug from the carriers or protecting groups. In this part, biomedical and material engineering applications of photocleavable compounds will be discussed.

2.4.3.1 Photocages

The term "caging" was coined in 1978 by J. F. Hoffman. It refers to the coupling of a therapeutic agent with a photoactive compound. The drug stays inactive until it is cleaved from the cage by light illumination. This process has great potentials in the controlled and targeted delivery of drugs. The most commonly used photocages are ONB, coumarin-4-yl methyl, *p*-hydroxyphenacyl, 7-nitroindoline, and 3′,5′-dialkoxybenzoin derivatives. Mayer and Heckel discussed the applications of photocages for biologically active molecules.[43] A few distinctive examples will be discussed, especially of the application methodologies of these types of materials.

Choi et al. reported the release of methotrexate (MTX) from a fifth-generation polyamidoamine (PAMAM) dendrimer covalently linked to MTX by an ONB linker.[44] First they coupled the carboxylic acid part of methotraxane with the benzyl alcohol part of ONB and attached this compound onto the PAMAM dendrimer. Additionally, folic acid was introduced onto the dendrimers for recognition of cancer cells (Fig. 2.19). The other advantage of the system is deactivation/activation of the drug. Since MTX is highly toxic, it must be activated only in cancer cells. When it is linked to the carrier, it is not active, and it doesn't show any toxic effect. Upon UV irradiation, it is activated and cancer cells are destroyed. This process is called "photocaging." The authors just showed the phototriggered release of MTX from the PAMAM dendrimer and the controlled release profile by high performance liquid chromatography (HPLC). Since there was no report of in vivo studies, the effectiveness of the system is not known.

Rotello and coworkers reported the release of caged fluorouracil from gold nanoparticles.[45] The toxic fluorouracil compound was linked to gold nanoparticles by an ONB linker (AU-ONB-FU) and upon irradiation with light of 365 nm the drug was released from the carrier. The small size of the carriers (10 nm) ensured long circulation time and accumulation in cancer tissues due to enhanced permeability and retention effect. The authors also showed in vivo studies of the systems. Control experiments indicated that the particles without UV irradiation were not toxic but UV light shows some phototoxicity. When particles were

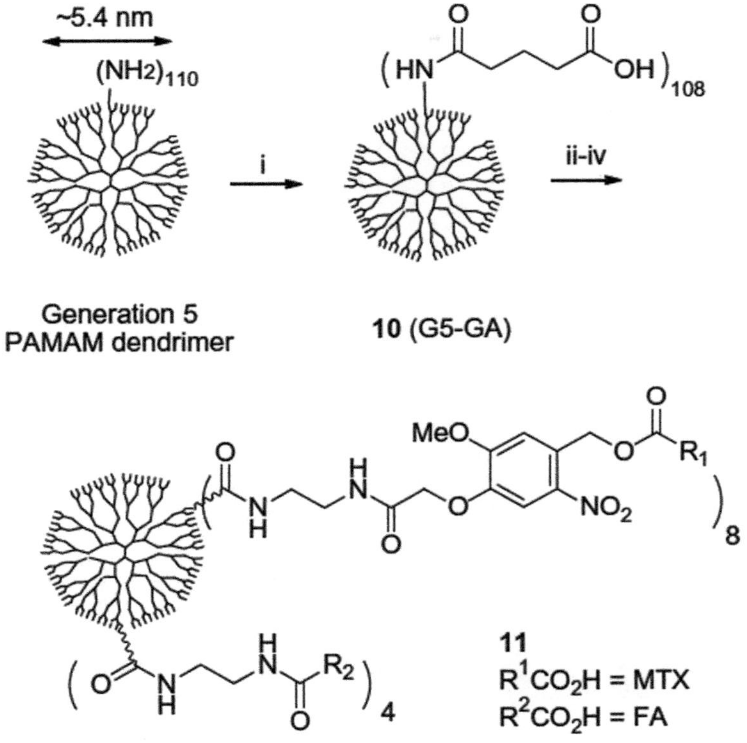

Figure 2.19 Synthesis of a folate receptor–targeting PAMAM dendrimer conjugated with methotrexate. Adapted from Ref. [44].

introduced into the cell, subsequent UV light irradiation caused remarkable cell death (Fig. 2.20).

As reported above, the main problem for drug delivery applications using photoresponsive systems is the phototoxicity of the UV light. In addition, the penetration depth of UV light through the skin and tissues is very limited. One way to overcome this problem is to use an upconversion process.[46] Branda and coworkers showed a nice combination of UCPs with photocages.[47] They

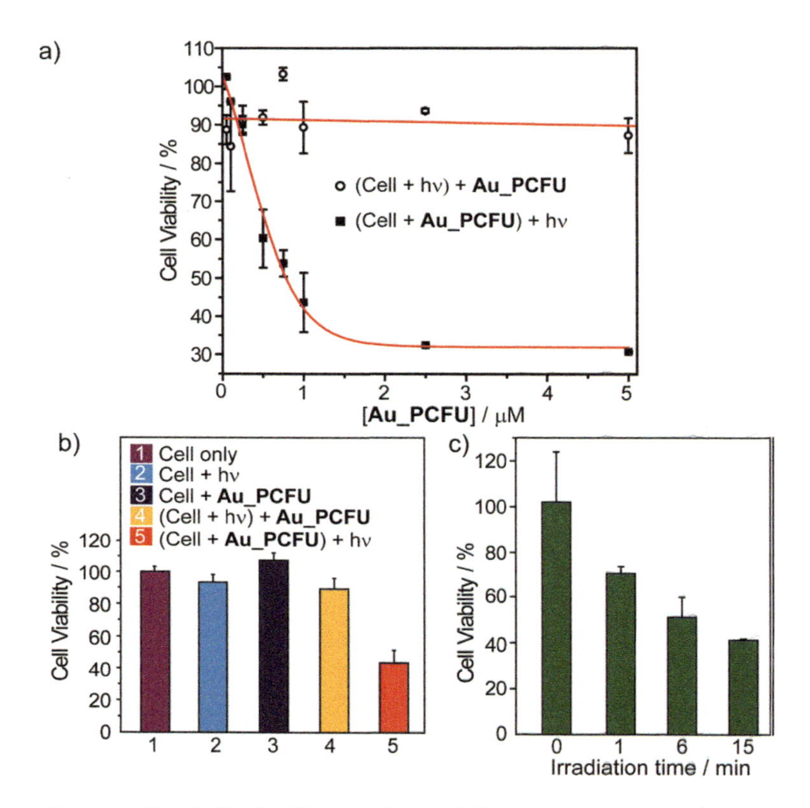

Figure 2.20 Cell viability studies of AU-ONB-FU. Adapted from Ref. [45].

immobilized 3′,5′-dialkoxybenzoin onto UCPs and acetic acid as a generic molecule for caging. They showed photocleavage and release of acetic acid from benzoin upon irradiation with light of 980 nm.

2.4.3.2 Self-assembly

Photocleavable compounds can be used for drug release from micelles. Since micelles contain hydrophobic and hydrophilic segments, photocleavable moieties can be used as a trigger to change this balance. For instance, Liu et al. used ONB-coupled cysteine as a hydrophobic monomer.[48] They synthesized poly(*S*-(*o*-nitrobenzyl)-L-cysteine)-*b*-poly(ethylene oxide). Upon UV irradiation, the ONB part was separated, causing a change in the hydrophilic-hydrophobic balance of the micelles. This structural change resulted in the release of loaded doxorubicin from the micelles.

Branda and coworkers used a similar system with UCPs. They polymerized ONB–bearing methacrylate (PONBMA) by using a PEO-based macroinitiator for ATRP.[49] The resulting amphiphilic block copolymer (PEO-*b*-PONBMA) was used for micellization and was loaded with UCPs. Infrared light (980 nm) was upconverted to light of 365 nm by UCPs and resulted in the photocleavage of ONB groups. Since the remaining methacrylic acid units are hydrophilic, the micelles were destroyed and loaded UCPs and hydrophobic payload (Nile red) were released. The same group extended their study to hydrogel-based drug delivery.[50] They loaded UCPs and trypsin into a hydrogel that contained an ONB group–bearing cross-linker. Trypsin was released from the hydrogels by near-infrared-light-irradiation-triggered dissociation of the cross-linkers.

Meier and coworkers used ONB molecules in the middle of hydrophobic and hydrophilic segments.[51] They functionalized ONB with a halide group. First, they synthesized poly(methyl caprolactone)-*o*-nitrobenzyl by ring-opening polymerization of α-methyl-ε-caprolactone, which was initiated from the hydroxyl part of the ONB. The subsequent atom-transfer radical polymerization of *t*-butyl acrylate from the halide group and hydrolysis to acrylic acid

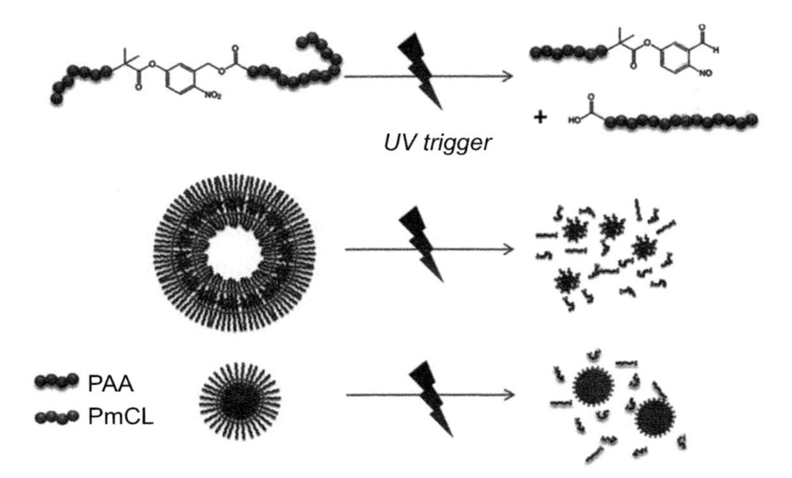

Figure 2.21 Phototriggered release of doxorubicin. Adapted from Ref. [51].

results in amphiphilic, biodegradable, and photocleavable block copolymer (Fig. 2.21).

Woodcock et al. reported the synthesis of thermo- and photoresponsive aqueous gels from thermo- and light-sensitive hydrophilic ABA triblock copolymers.[52] They used a PEO-based difunctional ATRP initiator and synthesized poly(ethoxytri(ethylene glycol) acrylate-*co*-*o*-nitrobenzyl acrylate)-*b*-poly(ethylene oxide)-*b*-poly (ethoxytri(ethylene glycol) acrylate-*co*-*o*-nitrobenzyl acrylate) polymer, or P(TEGEA-*co*-NBA)-*b*-PEO-*b*-P(TEGEA-*co*-NBA) polymer, which is thermo- and photoresponsive (Fig. 2.22). An aqueous solution of the polymer showed sol-gel behavior upon heating. Above 42°C the polymer solution reversibly turned to gel totally. After UV light irradiation, the critical temperature shifted to 53°C due to the more hydrophilic character of acrylic acid, which is left after cleavage of the ONB part.

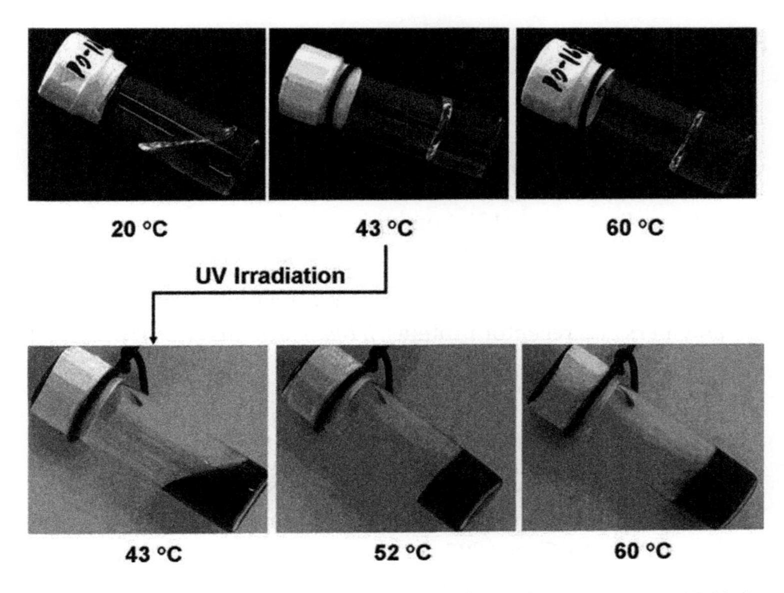

Figure 2.22 Photo- and thermoresponsivity of P(TEGEA-*co*-NBA)-*b*-PEO-*b*-P(TEGEA-*co*-NBA). Adapted from Ref. [52].

Han et al. synthesized multiresponsive ABA-type amphiphilic block copolymers with redox or photocleavable units in their main chain.[53] Two propargyl groups bearing disulfide-based monomers and two azide groups bearing ONB-based monomers were polymerized by click reaction–based condensation polymerization. The authors kept the amount of propargyl-bearing monomer slightly higher to obtain propargyl groups as the end functionality. Then, they attached PEO-N_3 units and obtained an ABA-type amphiphilic block copolymer. The obtained Nile red–doped micelles were used to control the stimuli-responsive release behavior of the system. Han et al. showed that the micelles released their cargo upon treatment with UV light or the reducing agent dithiothreitol.

References

1. Hoffman, A. S. (2013). Stimuli-responsive polymers: biomedical applications and challenges for clinical translation, *Adv. Drug Delivery Rev.*, **65**(1), pp. 10–16.
2. Cabane, E., et al. (2012). Stimuli-responsive polymers and their applications in nanomedicine, *Biointerphases*, **7**(1).
3. Hoffman, A. S., et al. (2000). Really smart bioconjugates of smart polymers and receptor proteins, *J. Biomed. Mater. Res.*, **52**(4), pp. 577–586.
4. Qiu, Y. and Park, K. (2012). Environment-sensitive hydrogels for drug delivery, *Adv. Drug Delivery Rev.*, **64**, pp. 49–60.
5. Jeong, B. and Gutowska, A. (2002). Lessons from nature: stimuli-responsive polymers and their biomedical applications, *Trends Biotechnol.*, **20**(7), pp. 305–311.
6. Kobayashi, J., et al. (2002). Aqueous chromatography utilizing hydrophobicity-modified anionic temperature-responsive hydrogel for stationary phases, *J. Chromatogr. A*, **958**(1–2), pp. 109–119.
7. Peng, T. and Cheng, Y. L. (2001). PNIPAAm and PMAA co-grafted porous PE membranes: living radical co-grafting mechanism and multi-stimuli responsive permeability, *Polymer*, **42**(5), pp. 2091–2100.
8. Gibson, M. I. and O'Reilly, R. K. (2013). To aggregate, or not to aggregate? considerations in the design and application of polymeric thermally-responsive nanoparticles, *Chem. Soc. Rev.*, **42**(17), pp. 7204–7213.
9. Zhang, K. and Khan, A. (1995). Phase behavior of poly(ethylene oxide)-poly(propylene oxide)-poly(ethylene oxide) triblock copolymers in water, *Macromolecules*, **28**(11), pp. 3807–3812.
10. Liu, S. and Armes, S. P. (2001). The facile one-pot synthesis of shell cross-linked micelles in aqueous solution at high solids, *J. Am. Chem. Soc.*, **123**(40), pp. 9910–9911.

11. Alarcon, C. d. l. H., Pennadam, S. and Alexander, C. (2005). Stimuli responsive polymers for biomedical applications, *Chem. Soc. Rev.*, **34**(3), pp. 276–285.

12. Kaneko, Y., et al. (1996). Fast swelling/deswelling kinetics of comb-type grafted poly(N-isopropylacrylamide) hydrogels, *Macromol. Symp.*, **109**(1), pp. 41–53.

13. Gandhi, A., et al. (2015). Studies on thermoresponsive polymers: phase behaviour, drug delivery and biomedical applications, *Asian J. Pharm. Sci.*, **10**(2), pp. 99–107.

14. Roy, D., Brooks, W. L. A. and Sumerlin, B. S. (2013). New directions in thermoresponsive polymers, *Chem. Soc. Rev.*, **42**(17), pp. 7214–7243.

15. Bradley, C., et al. (2009). Response characteristics of thermoresponsive polymers using nanomechanical cantilever sensors, *Macromol. Chem. Phys.*, **210**(16), pp. 1339–1345.

16. Lavigne, M. D., et al. (2007). Enhanced gene expression through temperature profile-induced variations in molecular architecture of thermoresponsive polymer vectors, *J. Gene Med.*, **9**(1), pp. 44–54.

17. Mao, Z., et al. (2007). The gene transfection efficiency of thermoresponsive N,N,N-trimethyl chitosan chloride-g-poly (N-isopropylacrylamide) copolymer, *Biomaterials*, **28**(30), pp. 4488–4500.

18. Siegel, R. A., et al. (1988). pH-controlled release from hydrophobic/polyelectrolyte copolymer hydrogels, *J. Control. Release*, **8**(2), pp. 179–182.

19. Ghandehari, H., Kopečková, P. and Kopecek, J. (1997). In vitro degradation of pH-sensitive hydrogels containing aromatic azo bonds, *Biomaterials*, **18**(12), pp. 861–872.

20. Gao, W., Chan, J. M. and Farokhzad, O. C. (2010). pH-responsive nanoparticles for drug delivery, *Mol. Pharmaceutics*, **7**(6), pp. 1913–1920.

21. Kim, J.-w. and Dang, C. V. (2006). Cancer's molecular sweet tooth and the warburg effect, *Cancer Res.*, **66**(18), pp. 8927–8930.

22. Griset, A. P., et al. (2009). Expansile nanoparticles: synthesis, characterization, and in vivo efficacy of an acid-responsive polymeric drug delivery system, *J. Am. Chem. Soc.*, **131**(7), pp. 2469–2471.

23. Roy, D., Cambre, J. N. and Sumerlin, B. S. (2010). Future perspectives and recent advances in stimuli-responsive materials, *Prog. Polym. Sci.*, **35**(1–2), pp. 278–301.

24. Peng, K., Tomatsu, I. and Kros, A. (2010). Light controlled protein release from a supramolecular hydrogel, *Chem. Commun.*, **46**(23), pp. 4094–4096.

25. Patnaik, S., et al. (2007). Photoregulation of drug release in azo-dextran nanogels, *Int. J. Pharm.*, **342**(1–2), pp. 184–193.

26. Guo, W., et al. (2014). Triple stimuli-responsive amphiphilic glycopolymer, *J. Polym. Sci., Part A: Polym. Chem.*, **52**(15), pp. 2131–2138.

27. Lu, J., et al. (2008). Light-activated nanoimpeller-controlled drug release in cancer cells, *Small*, **4**(4), pp. 421–426.

28. Croissant, J., et al. (2013). Two-photon-triggered drug delivery in cancer cells using nanoimpellers, *Angew. Chem. Int. Ed.*, **52**(51), pp. 13813–13817.

29. Qiu, Z., et al. (2009). Spiropyran-linked dipeptide forms supramolecular hydrogel with dual responses to light and to ligand-receptor interaction, *Chem. Commun.*, **2009**(23), pp. 3342–3344.

30. Lee, H.-i., et al. (2007). Light-induced reversible formation of polymeric micelles, *Angew. Chem. Int. Ed.*, **46**(14), pp. 2453–2457.

31. Zhou, Y.-N., Zhang, Q. and Luo, Z.-H. (2014). A light and pH dual-stimuli-responsive block copolymer synthesized by copper(0)-mediated living radical polymerization: solvatochromic, isomerization, and "schizophrenic" behaviors, *Langmuir*, **30**(6), pp. 1489–1499.

32. Jin, Q., Liu, G. and Ji, J. (2010). Micelles and reverse micelles with a photo and thermo double-responsive block copolymer, *J. Polym. Sci., Part A: Polym. Chem.*, **48**(13), pp. 2855–2861.

33. Feng, K., et al. (2012). Reversible light-triggered transition of amphiphilic random copolymers, *Macromolecules*, **45**(13), pp. 5596–5603.

34. Trenor, S. R., et al. (2004). Coumarins in polymers: from light harvesting to photo-cross-linkable tissue scaffolds, *Chem. Rev.*, **104**(6), pp. 3059–3077.

35. Wells, L. A., Brook, M. A. and Sheardown, H. (2011). Generic, anthracene-based hydrogel crosslinkers for photo-controllable drug delivery, *Macromol. Biosci.*, **11**(7), pp. 988–998.

36. Zheng, Y., et al. (2002). PEG-based hydrogel synthesis via the photodimerization of anthracene groups, *Macromolecules*, **35**(13), pp. 5228–5234.

37. Shi, D., Matsusaki, M. and Akashi, M. (2009). Photo-cross-linking induces size change and stealth properties of water-dispersible cinnamic acid derivative nanoparticles, *Bioconjugate Chem.*, **20**(10), pp. 1917–1923.

38. He, J., Tong, X. and Zhao, Y. (2009). Photoresponsive nanogels based on photocontrollable cross-links, *Macromolecules*, **42**(13), pp. 4845–4852.

39. Chung, C.-M., et al. (2004). Crack healing in polymeric materials via photochemical [2+2] cycloaddition, *Chem. Mater.*, **16**(21), pp. 3982–3984.

40. Bochet, C. G. (2002). Photolabile protecting groups and linkers, *J. Chem. Soc., Perkin Trans. 1*, (2), pp. 125–142.

41. Chandra, B., et al. (2006). Formulation of photocleavable liposomes and the mechanism of their content release, *Org. Biomol. Chem.*, **4**(9), pp. 1730–1740.

42. Kim, M. S. and Diamond, S. L. (2006). Photocleavage of o-nitrobenzyl ether derivatives for rapid biomedical release applications, *Bioorg. Med. Chem. Lett.*, **16**(15), pp. 4007–4010.

43. Mayer, G. and Heckel, A. (2006). Biologically active molecules with a "light switch," *Angew. Chem. Int. Ed.*, **45**(30), pp. 4900–4921.

44. Choi, S. K., et al. (2012). A photochemical approach for controlled drug release in targeted drug delivery, *Bioorg. Med. Chem.*, **20**(3), pp. 1281–1290.

45. Agasti, S. S., et al. (2009). Photoregulated release of caged anticancer drugs from gold nanoparticles, *J. Am. Chem. Soc.*, **131**(16), pp. 5728–5729.

46. Wang, X., et al. (2012). ortho-Nitrobenzyl alcohol based two-photon excitation controlled drug release system, *RSC Adv.*, **2**(1), pp. 156–160.

47. Carling, C.-J., et al. (2010). Remote-control photorelease of caged compounds using near-infrared light and upconverting nanoparticles, *Angew. Chem. Int. Ed.*, **49**(22), pp. 3782–3785.

48. Liu, G. and Dong, C.-M. (2012). Photoresponsive poly(S-(o-nitrobenzyl)-l-cysteine)-b-PEO from a l-cysteine N-carboxyanhydride monomer: synthesis, self-assembly, and phototriggered drug release, *Biomacromolecules*, **13**(5), pp. 1573–1583.

49. Yan, B., et al. (2011). Near-infrared light-triggered dissociation of block copolymer micelles using upconverting nanoparticles, *J. Am. Chem. Soc.*, **133**(49), pp. 19714–19717.

50. Yan, B., et al. (2012). Near infrared light triggered release of biomacromolecules from hydrogels loaded with upconversion nanoparticles, *J. Am. Chem. Soc.*, **134**(40), pp. 16558–16561.

51. Cabane, E., Malinova, V. and Meier, W. (2010). Synthesis of photocleavable amphiphilic block copolymers: toward the design of photosensitive nanocarriers, *Macromol. Chem. Phys.*, **211**(17), pp. 1847–1856.

52. Woodcock, J. W., et al. (2010). Dually responsive aqueous gels from thermo- and light-sensitive hydrophilic ABA triblock copolymers, *Soft Matter*, **6**(14), pp. 3325–3336.

53. Han, D., Tong, X. and Zhao, Y. (2012). Block copolymer micelles with a dual-stimuli-responsive core for fast or slow degradation, *Langmuir*, **28**(5), pp. 2327–2331.

Chapter 3

Stimuli-Responsive Polymers with Tunable Release Kinetics

Mehmet Can Zeybek,* **Egemen Acar, and Gozde Ozaydin-Ince**
Materials Science and Nanoengineering Program, Faculty of Engineering and Natural Sciences, Sabanci University, 34956 Istanbul, Turkey
gozdeince@sabanciuniv.edu

3.1 Introduction

Polymeric materials are one of the fastest-growing classes of materials due to their easy processability, cost effectiveness, and convenience in terms of a wide range of applications. Stimuli-responsive (smart) polymers are polymers with the capability of responding to several physical and chemical environmental stimuli. Slight changes in the external stimuli of light, electrical current, temperature, pH, or glucose concentration trigger physicochemical changes

*Current address: ARC Centre of Excellence in Convergent Bio-Nano Science and Technology and Department of Chemical Engineering, The University of Melbourne, Parkville, Victoria 3010, Australia.

Switchable Bioelectronics
Edited by Onur Parlak

ISBN 978-981-4800-89-1 (Hardcover), 978-1-003-05600-3 (eBook)
www.jennystanford.com

(such as shape change, variation in color, or change in conductivity) in the polymeric materials. The ability to control the response of these materials externally has prompted their use in a wide range of applications, including microfluidics, textiles, sensors, packaging, aerospace, and biotechnology.[46]

The term "tunable release" refers to the release of certain molecules in a medium with precise control over the release quantity and the rate of release in a predefined time range. The type of release method to be employed in a given application depends on the requirements, such as zero payload waste, a constant release rate for a period of time, or suppressed negative effects of therapeutic molecules, as in the case of drug delivery.[7] The main types of release mechanisms that are commonly used are:

- Immediate release: The payload is directly released as it enters the targeted medium.
- Delayed release: The payload is released with an intentional delay in the onset of release.
- Burst release: The payload is released rapidly once the release is triggered and dissolves in the medium without any time delay.
- Sustained release: The payload is released over a period of time at a controlled rate once the release is triggered.
- Targeted release: Release takes place upon encounter of the carrier molecule with the targeted molecule.
- Pulsatile release: With the use of a preprogrammed system or periodic external stimuli, the payload is released at specified times.

In a given application, a single or a combination of these release mechanisms can be utilized depending on

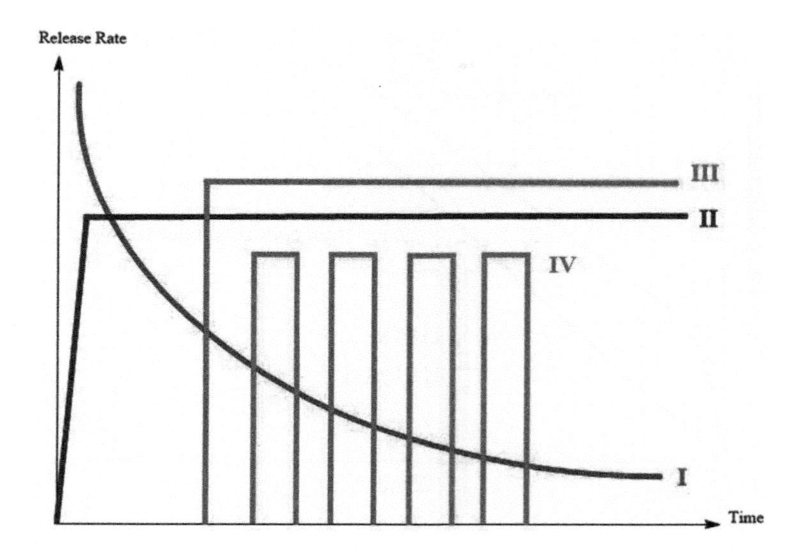

Figure 3.1 Common controlled release profiles used in biomedical applications.[90]

the release requirements. Controlled release refers to a combination of two or more of these release mechanisms to achieve prolonged release of molecules at specific release rates. Controlled release profiles that are commonly used in biomedical applications are shown in Fig. 3.1.[90] In Profile I, delayed release with varying release concentrations can be seen. Profile II depicts zero-order release, or immediate and sustained release, where the initial release concentration is maintained. This profile is considered as the ideal way for the release of antibiotics, antidepressants, and drugs for blood pressure maintenance and pain control.[74] Profile III shows a combination of delayed and sustained release paths. This method is used particularly for releasing active agents that need to be released during night in clinical applications.[90] Profile IV describes pulsatile release path after a delay period.[7]

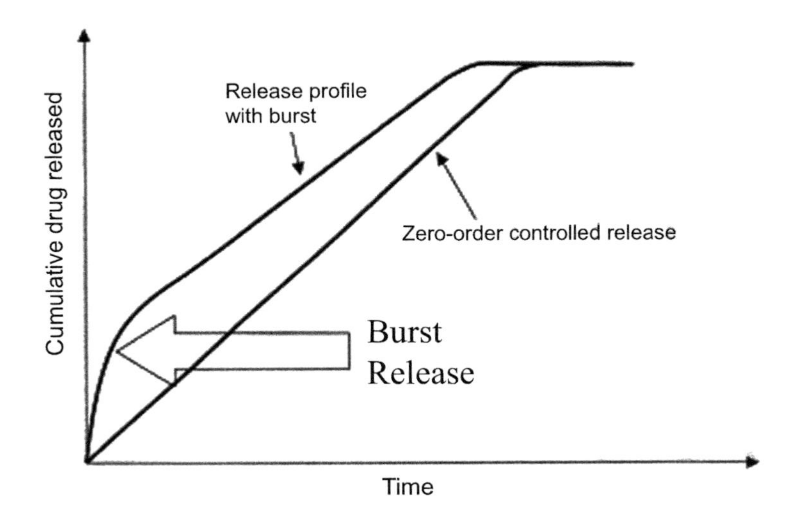

Figure 3.2 Burst release behavior in a drug release profile.[34]

In many of the release profiles, burst release is observed as the initial rapid release of high concentrations, as shown in Fig. 3.2.[34] Burst release is generally not desired due to the potential local toxicity and decreased release capacity and is economically wasteful. However, in certain applications, such as targeted release, wound treatment, or pulsatile release, it is significantly advantageous.[34]

Application requirements of the release mechanism determine whether the burst release systems are appropriate for the given application. Studies that benefit from burst release and focus on its prevention will be covered in examples given further in the chapter.

Besides the fact that tunable release systems have been widely used in biomedical studies, some of which will be detailed later as a part of this chapter, these systems are also regularly used in agriculture, food, and cosmetic industries. In food industries, for example, tunable release of relatively high molecular weight polysaccharide molecules can be exploited to enhance and preserve the food quality by prolonging the release of these molecules, and this concept is well explained in the literature.[91] In the cosmetics industry, especially in perfumery, many different specially designed fragrances are released in a tunable manner by utilizing different encapsulation techniques specific to the complex organic molecules used.[89] Another example is the controlled release of glycerol, which is the moisturizing agent in creams, over a period of time.[80]

In agriculture, as an important part of fertilizer management, tunable release systems are used extensively to ensure the maintenance of critical levels of fertilizers for long periods of time in fragile weather conditions.[43] Tunable release systems are most widely used in drug delivery research, which requires delicate control of the release, thus encouraging further studies on release kinetics. Therefore, a large number of the examples given on release kinetics will be related to drug delivery applications.

Drug delivery basically requires a material that will hold a sufficient amount of the therapeutic molecules within, retain them while moving to the target region, and release the therapeutic molecules in sufficient amounts across a desired time period.[82] It is essential to keep the levels of the therapeutic molecules within a treatment range to avoid side effects and prevent the patient from consuming high amounts of the drug, as shown in Fig. 3.3.[38] The free drug kinetics curve as shown in the figure represents the immediate release path of the drug on conventional administration. Each dosage causes fluctuations, involving a sudden increase and then a sudden decrease in drug concentration in the patient's blood, which is not desirable.

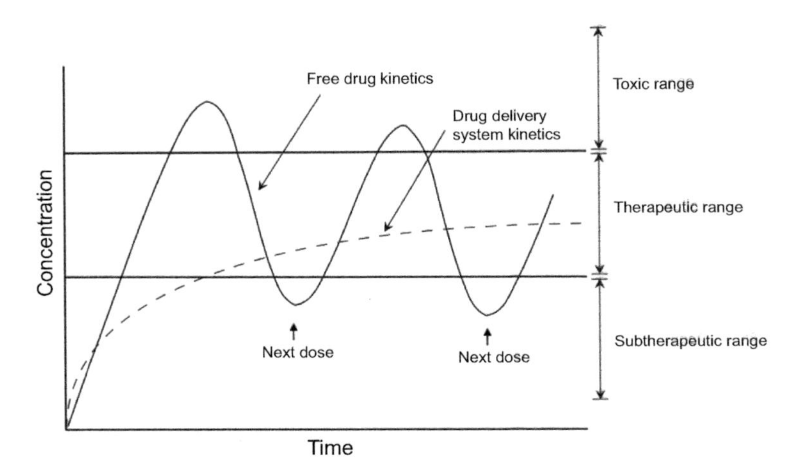

Figure 3.3 Release profile comparison of free drug kinetics and drug delivery system kinetics.[38]

For these reasons new ways of drug administration that reduce oscillations while keeping the drug concentrations below the toxic and above the subtherapeutic range should be developed.

Early controlled release studies used silicon tubes to entrap and release large molecules.[19] In this structure, the drug could diffuse along a tortuous and porous network. The studies that followed mainly focused on macromolecule releases. As a result of these studies, drug-containing hydrogels, biodegradable container systems, and matrices that allow the control of diffusion rates were developed.

The most commonly used drug delivery systems with different release mechanisms are shown in Fig. 3.4.[82] For matrix-based approaches, the diffusion of drug molecules through a network with an interconnected porous structure is the main delivery method. In erodible systems,

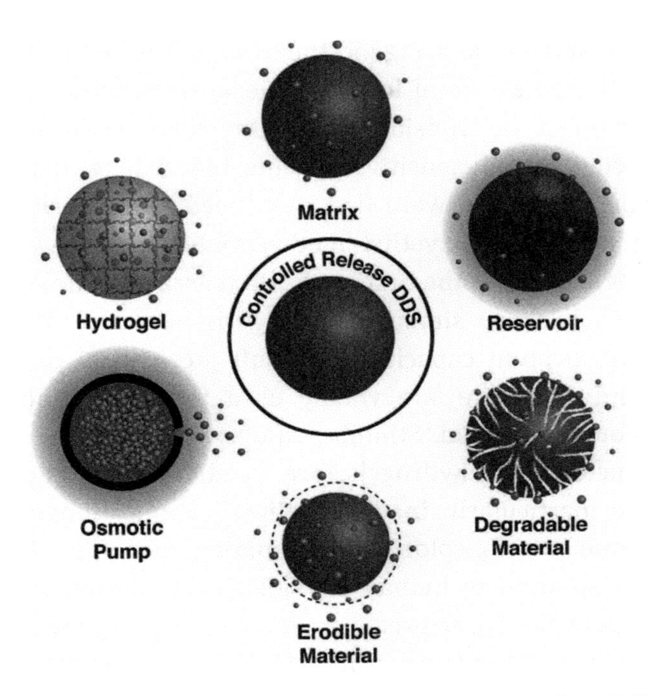

Figure 3.4 Various controlled release systems used as drug delivery platforms.[82]

intentional dissolution of the carrier matrix in a medium is exploited to control the drug release. Similarly, all along the degradation of the carrier material, the payload is released in the degradable material–based approach. In the reservoir systems, a membrane is used to let the drug diffuse out in desired amounts, which can be controlled by tuning the membrane properties. On the other hand, osmotic pump–based release systems use membranes impermeable to the drugs and benefit from the osmotic pressure differences to release the drugs. The big breakthrough in this field was the integration of hydrogels in the delivery systems. In the hydrogel-based controlled delivery systems a change in the structure of the polymer initiates the release of loaded molecules. Hydrogel-based systems can be considered as the early examples of advancing stimuli-sensitive-polymer-based drug delivery systems. Hydrogels were first developed around 1960 by using poly(2-hydroxyethyl methacrylate) (pHEMA), and contact lenses was the first application area of hydrogels.[84]

Hydrogels are similar to living tissues due to their large absorption capacities and soft mechanical properties and, therefore, are widely used for applications in the biomaterial field. Stimuli-responsive polymers, which also include the hydrogels, are candidates for tunable release, particularly in drug delivery applications. Ease of manufacturing, coloring and molding, biocompatibility, easy adaptation to human physiological conditions, non-thrombogenic characteristics, and adhesive nature with the use of some ligands are a few of the main advantageous of these hydrogels.[3,51] A thorough understanding of the release kinetics of a given system is crucial for the control of the release profile. Each individual drug delivery platform exhibits a different release mechanism, and the release kinetics of different controlled release systems can be studied using various theoretical and empirical models. Some commonly used release models and mechanisms are listed in Table 3.1.

For the release mechanisms with kinetics that can be described using well-established formulations, some of the commonly used models are the zero-order, first-order, Higuchi, Korsemeyer–Peppas, and Hixson–Crowell models. Zero-order kinetics models are used to describe the release kinetics where the drug release rate from the delivery platform remains constant throughout the process and is

Table 3.1 Mathematical models for drug release mechanisms

Kinetic model	Mathematical relation	Release mechanism	Systems that follow the model
Zero order	$Q_1 = Q_0 + K_0 t$	Diffusion mechanism	Osmotic systems; transdermal systems
First order	$\log C = \log C_0 - Kt/2.303$	Fick's first law; diffusion mechanism	Water-soluble drugs in a porous matrix
Higuchi model	$f_t = K_H t^{1/2}$	Diffusion medium–based mechanism in Fick's first law	Diffusion matrix formulations
Korsmeyer–Peppas model	$f_t = at^n$	Semi-empirical model; diffusion-based mechanism	Swellable polymeric devices
Hixson model	$W_0{}^{1/3} - W_t{}^{1/3} = K_S t$	Erosion release mechanism	Erodible matrix formulations

independent of the drug concentration. Release kinetics of the osmotic and transdermal systems and systems with drugs of low solubility can be studied using the zero-order kinetics models. The mathematical representation of this model is

$$Q_1 = Q_0 + K_0 t, \tag{3.1}$$

where Q_1 is the cumulative amount of the drug released at time t, Q_0 is the initial concentration of the drug, and K_0 is the release constant at zero order. This model is widely used for the prediction and control of the release behavior of antibiotics delivery systems, sustaining levels of blood and heart pressure, and pain control.[74]

First-order kinetics, on the other hand, is used to model the release from delivery systems in which the drug release rate is proportional to the amount of drug remaining. The mathematical relation describing this model is given as

$$\log C = \log C_0 - \frac{Kt}{2.303} \tag{3.2}$$

where C_0 is the initial concentration of the drug, C is the concentration of the drug remaining in the media, and the t is the process time. This model is generally used to analyze elimination or absorption mechanisms of certain drugs or the dissolution behavior of water-soluble drugs in porous matrices.[9]

The release behavior of some drugs from matrices of porous systems possessing different geometries is generally studied using the Higuchi model, which is given as

$$f_{\mathrm{t}} = K_{\mathrm{H}} t^{1/2}, \tag{3.3}$$

where f_t is the fraction of the dissolved drug at time t and K_{H} is the Higuchi dissolution constant. The Higuchi model takes Fick's law as the basis to model the diffusion mechanism during drug release and is applied to analyze the release behavior of drugs at different dosages in certain transdermal systems. Korsmeyer et al.[44] proposed a simple formulation to model the release of drugs from polymeric systems, in which the time dependence of drug release is in the exponential form, given by

$$f_{\mathrm{t}} = at^n \tag{3.4}$$

where f_t is the fraction of drug released at time t, a is the rate constant for the release process, and n is the exponential constant of the release, which is used to analyze release processes for matrices and is an indicant of the drug release mechanism.[44] Peppas et al. further studied different drug release mechanisms and for each transport mechanism, defined a range of n values for which the given mechanism is applicable as listed in Table 3.2.[63]

The Korsemeyer–Peppas model is generally applied to examine the release studies if the mechanism of drug release is not well known. Another widely used model to study the kinetics of drug release is the Hixson–Crowell model, which describes the controlled release from systems with dimensions and surface areas that change throughout the process. Using the fact that the surface area of the

Table 3.2 Different diffusion release mechanisms based on the value of the exponential constant n

Release exponent (n)	Drug transport mechanism	Rate as a function of time
0.5	Fickian diffusion	$t^{-0.5}$
$0.45 < n = 0.89$	Non-Fickian transport	t^{n-1}
0.89	Case II transport	Zero-order release
Higher than 0.89	Super case II transport	t^{n-1}

particle and the cubic root of the particle volume have a proportional relation, the following equation is derived:

$$W_0^{1/3} - W_t^{1/3} = K_s t, \qquad (3.5)$$

where W_0 is the initial amount of the drug, W_t is the amount of the drug at time t, and K_s is the constant that embodies the relation between the surface and the volume.[31] The drug release systems that basically follow this model are the erodible matrix systems with drug particles that have a decreasing surface area and volume while they retain the geometrical form of the initial state during the process.[15]

3.2 Stimuli-Responsive Polymers

Stimuli-responsive polymers can be categorized according to the types of stimuli that trigger their physicochemical responses. In this section, the most widely used stimuli-responsive polymers, their usage in controlled delivery applications, and their release kinetics will be discussed.

3.2.1 Temperature-Responsive Polymers

Smart polymers with both hydrophobic and hydrophilic groups in their chemical structure are sensitive to environmental stimuli due to the sensitive balance between these functional groups. Temperature-sensitive polymers are a group of smart polymers that undergo a physicochemical change when exposed to a slight change in the temperature of their environment.[36]

A temperature-responsive behavior is particularly useful for biomedical applications because temperature change is a major and primitive response of the human body during local infections and diseases. Therefore, temperature-responsive polymers with different structures, such as particles, interpenetrating networks, thin films, and micelles, are suitable for a wide range of applications in the biomedical field. Besides, considerable effort is being made in developing smart textiles[16] and performing bioseparation[32] with the use of temperature-sensitive polymers.

First, a thorough understanding of the basic phenomena behind the thermal-responsive behavior and the relation between the structure and properties of the polymer is required. Two dissimilar types of thermal behaviors of polymers in a solution have been observed and described using the phase diagrams shown in Fig. 3.5.[94] Increasing temperature causes some polymers to become hydrophobic and insoluble above a certain temperature. In such solutions on increasing the temperature, there is a failure to maintain the balance between the hydrophobic and hydrophilic sites. The weakening of the intermolecular hydrogen interactions between the polymer and water

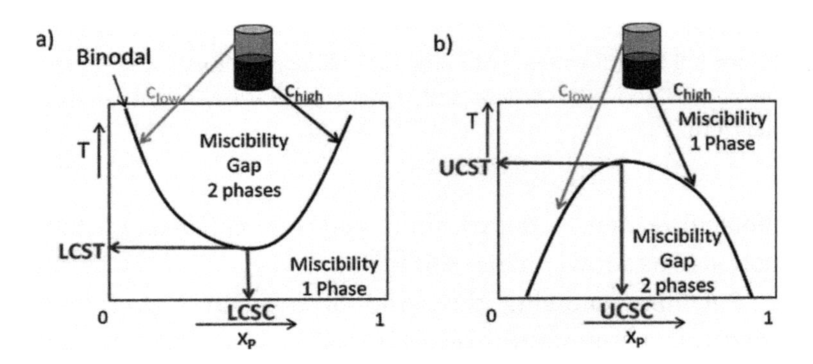

Figure 3.5 Phase diagrams of two different types of temperature-sensitive polymers in a solution.[94]

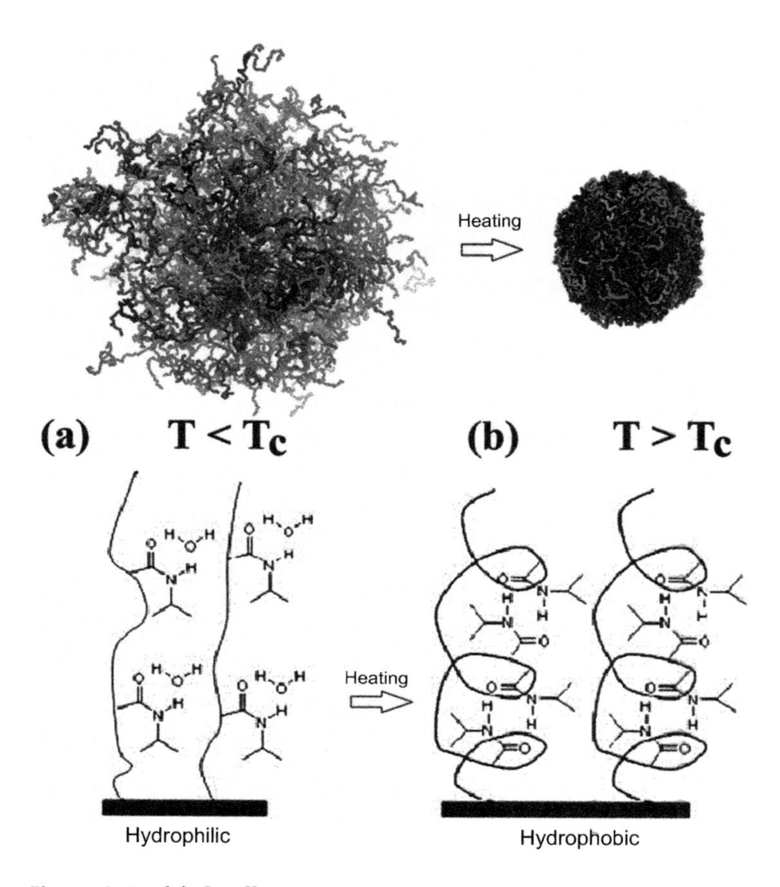

Figure 3.6 (a) Swollen state and (b) collapsed state of a polymer below and above the critical transition temperature. Adapted from Refs. [48, 55].

molecules causes dehydration and the collapse of the polymer, as shown in Fig. 3.6.[48,55]

Polymers showing this kind of behavior are called "negative temperature-sensitive systems" and the phase transition temperature is called the "lower critical solution temperature" (LCST). On the other hand, positive temperature-sensitive polymers collapse and show insoluble behavior upon cooling and the phase transition temperature in this case is called the "upper critical solution temperature" (UCST).

These phase transitions generally take place in a narrow temperature range and are reversible.[64] However, in the presence of sufficient H-bond donors in the molecular structure of polymers, inter- and intramolecular hydrogen bonds may hinder the rehydration upon cooling, causing hysteresis.[33] It must be noted that while for the solubility of positive temperature-sensitive systems short-range van der Waals interactions are important and they usually dissolve in organic solvents, for the solubility of negative temperature-sensitive systems, interactions of hydrogen bonds are dominant and they mostly dissolve in aqueous solvents.[21,33]

UCST-type polymers are considered unsuitable for biomedical applications due to their increasing solubility with increasing temperature. The aim of loading the injectable temperature-sensitive systems with the maximum amount of drug for the highest efficiency is that the drug and the carrier system should have high solubility under laboratory conditions. After the highly soluble carrier system is loaded with the drug, the drug-loaded delivery system is administered to the patient via a syringe. Inside the body, the delivery platform should have low solubility so that controlled release can take place at preplanned time scales. Since the human body temperature is higher than the temperature in laboratory conditions, temperature-responsive systems should have less solubility at high temperatures.[92] Consequently, most of the studies are focused on LCST-type polymers.[33] Poly(*N*-isopropyl acrylamide) (pNIPAAM), with 32°C LCST; poly(*N*-vinyl caprolactam), with 32°C–35°C LCST; poly(*N*,*N*-diethyl acrylamide); chitosan; poly(*N*-vinyl alkylamide); poly(lactide-*co*-glycolide)-poly(ethylene glycol)-poly(lactide-*co*-glycolide) (PLGA-PEG-PLGA); triblock copolymers; phosphazene; and polysaccharide derivatives can be given as examples of LCST-type temperature-responsive polymers.[13,69]

Temperature-responsive polymers are widely used in drug delivery applications, providing the benefit of triggering the drug release by controlling either the internal or the external stimuli. Some observations have revealed that a high metabolic activity rate, leukocyte infiltration, a high proliferation cell rate, and extraordinary blood flow directly lead to high body temperature in inflammatory types of diseases and tumors.[2,35,95] For the cases when the polymer stimulation should be controlled externally, conventional clinical methods, such as radiofrequency, ultrasound, and focused microwaves, can be applied for heat localization to create mild hyperthermia.[37,39,59]

Studies of temperature-sensitive polymers in drug delivery applications are mainly focused on the adjustment of the release kinetics, providing biocompatibility/biodegradability properties and enhancing mechanical strength. Recent studies have shown that optimization of the chain length between the hydrophobic and hydrophilic groups has a strong influence on the adjustment of release kinetics. Using temperature-responsive polymers, Choi et al. prepared a hydrogel system at 25°C and investigated its release characteristics at 37°C. This triblock copolymer system acted as a free-flowing sol and turned into a gel at 37°C, with a dramatically decreased flow. Moreover, the hydrogel system was injected subcutaneously and could retain its integrity for more than 2 weeks.[13] The hydrogel system developed revealed a huge burst release. More than 60% of the loaded drug was released within the first day, and the released quantity reached 80% in the next 3 days. On the other hand, the initial burst release of drugs could be slowed for more than 1 month by combining with a triblock PCL-PEG-PCL polymer system and stabilizing the drug structure with Zn complexation.[13] It was concluded in the study that the mechanism behind the prolonging was not the poor solubility of Zn complexed hydrogel but mainly the entrapment of the Zn complexed drug in the network of the triblock copolymer. With that work, the triblock copolymer system gained delayed release characteristics. As a result of this prolonged release, fewer drug injections were needed, increasing cost effectiveness and patient compliance.

Another study focusing on the release kinetics from polymeric systems is reported by Ince et al. In this study a coaxial nanotube structure was designed and synthesized using a templated chemical vapor deposition (CVD) method. The coaxial nanotube structures were composed of an outer layer of a shape memory polymer, p(TBA-*co*-DEGDVE), and an inner hydrogel layer of p(HEMA-*co*-EGDA). The combination of the inner hydrogel layer, which acted as a host for the fluorescent dye, and the temperature-responsive outer layer enabled the burst release of the model dye molecules, as depicted in Fig. 3.7.[61]

The temperature dependence of the outer layer is exploited to fix the diameter of the coaxial nanotubes, which can be changed by the swelling of the inner hydrogel layer. After loading the nanotubes with molecules by the swelling of the hydrogel layer, the release of the molecules is

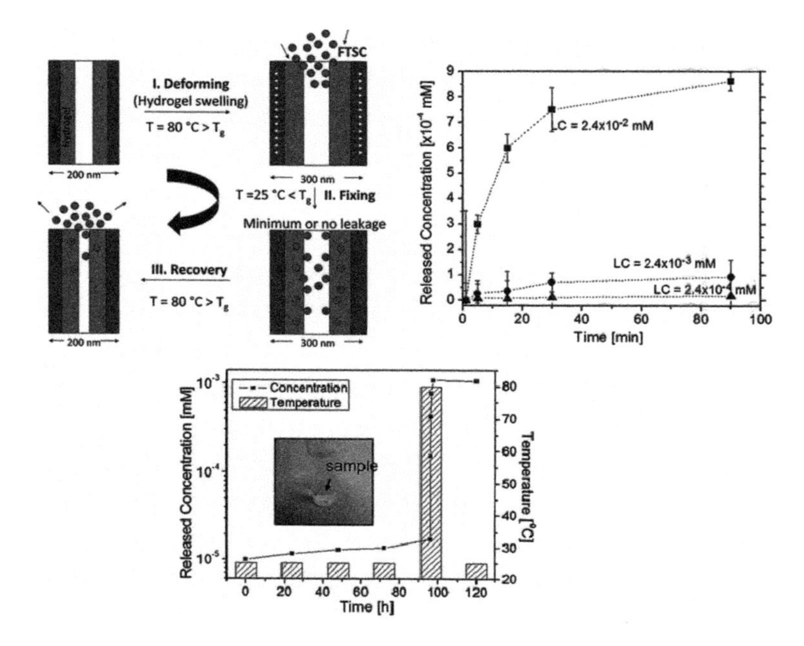

Figure 3.7 A novel molecular release system with loading and release schematics (left) and burst release profiles of nanotubes loaded with different dye concentrations (right).[61]

triggered by the rapid collapse of the outer shape memory polymer layer by increasing the temperature to a value above the transition temperature. This applied stress on the inner layer induces the burst release of the model molecules.[61]

Temperature-responsive drug delivery studies have focused mostly on utilizing pNIPAAM in the bulk polymer form or as nanospherical structures, triggering the release of the drug molecules by heating the medium. In a study by Shen et al. poly[*N*-(isopropyl acrylamide)-*co*-acrylamide-*co*-allylamine]-conjugated albumin nanospheres were prepared as carriers for the anticancer drug adriamycin.

The therapeutic agent was entrapped in the albumin nanospheres during preparation. The release of the drug from the carrier was triggered by heating the medium from 30°C to 43°C, and controlled release was achieved at the desired rates.[72]

In another work, Zhang et al. reported the development of an interpenetrating polymer network (IPN) structure of thermosensitive pNIPAAM hydrogel with the aim of achieving prolonged release and improved mechanical properties. Bovine serum albumin was chosen as the payload, and release studies were accomplished at 22°C and 37°C. It was noticed that the mechanical properties of IPNs could be improved considerably, and controlled release was observed with a fast response rate. However, during the first four hours, burst release was observed as well, though the overall release rate within the first ninety-six hours could be sustained at 28%.[88] With a high surface-to-volume ratio nanotubes have considerable advantages. However, due to some challenges in their production steps studies on pNIPAAM nanotubes are limited.[10,12,24]

Coaxial nanotubes with layers of polymers sensitive to different types of stimuli can be successfully used as delivery platforms. The ability to trigger the release using several different stimuli enables better tuning of the release kinetics, which is crucial for drug delivery applications. Armagan and Ince designed coaxial nanotubes of temperature-responsive pNIPAAM polymers with pH-responsive poly(methacrylic acid) (pMAA) and pHEMA polymer inner layers. The release kinetics of model molecules was studied under different pH and temperature conditions, and the loading and release conditions for the highest yield could be optimized. The graphs depicting the percentage of molecules released from different nanotube systems are shown in Fig. 3.8.[5] Each one of the release data is fitted to the following empirical model:

$$x(t) = x_{\mathrm{f}}(1 - e^{-kt}), \tag{3.6}$$

where k is the rate constant and x is the release percentage at time t.

The fastest release kinetics was observed for single-layer pNIPAAM nanotubes, whereas release from the coaxial nanotubes occurred at lower rates, providing better control over the kinetics, which is desirable for drug delivery applications.[5]

Another commonly used structure in tunable release studies is polymeric micelles, which are easy to synthesize with well-defined dimensions. In a different study conducted by Li et al.,[47] micelles of the thermoresponsive amphiphilic poly(methyl methacrylate)-*b*-poly(*N*-isopropyl acrylamide)-*b*-poly(methyl methacrylate) triblock copolymer were fabricated and their potential as drug carriers was investigated. In this delivery system, the hydrophobic drug is attached to the hydrophobic core while the hydrophilic shell protects the drug from external conditions and contains them until release. In this study, the low–water soluble anti-inflammatory drug prednisone acetate is used as the model drug and 22.5 wt% of the drug is loaded into the temperature-responsive triblock copolymer micelle system.

They reveal that the micelles maintain their structural form and the release is very low at temperatures below LCST, whereas above the LCST, the outer shells of the micelles become hydrophobic and the original structures can no longer be maintained. This structural change that

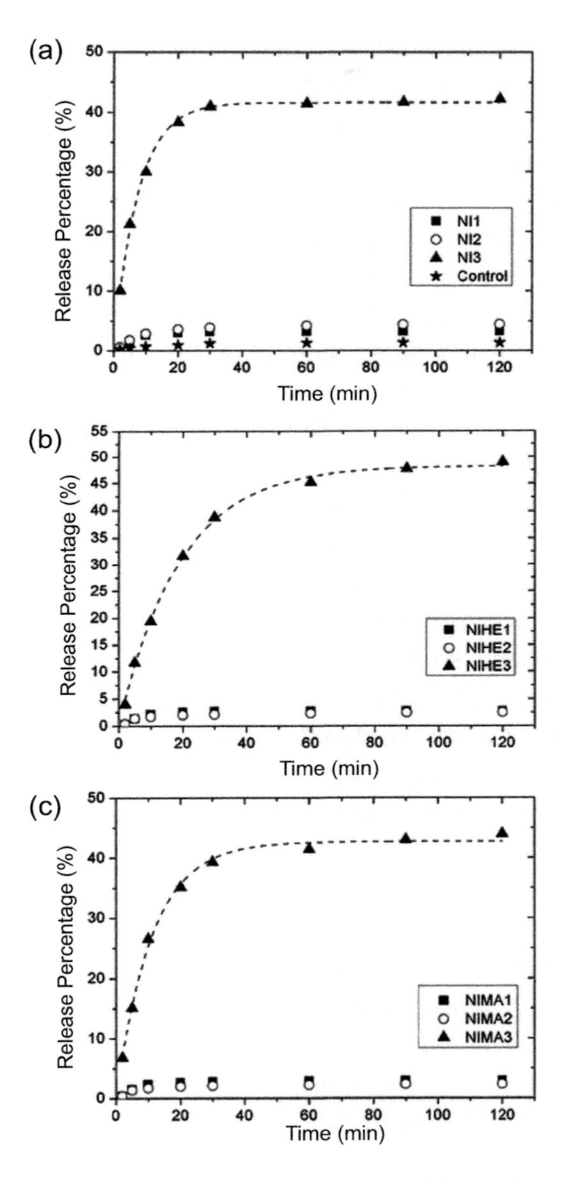

Figure 3.8 Drug release characteristics of (a) single pNIPAAM nanotubes, (b) coaxial pNIPAAM nanotubes with a pHEMA inner layer, and (c) coaxial pNIPAAM nanotubes with a pMAA inner layer.[5]

occurs above the LCST accelerates the release of drugs. In this study, sustained release of the drug could be achieved for 100 hours.[47]

For a clinically applicable drug delivery system, the temperature-responsive polymer, which contains the drug, should be both biodegradable and biocompatible. If the polymeric system is not biodegradable, after completing its drug delivery task, it should be excreted from the patient's body to prevent toxic side effects.[40] Another concern for the drug delivery systems is the biocompatibility of the polymers used. Although pNIPAAM polymer is a good candidate for drug delivery systems, due to the unique sharp and almost discontinuous phase transition and an LCST near the human body temperature, having quaternary ammonium in the chemical structure makes the polymer cytotoxic and limits its usage in the biomedical field.[67] Recent studies confirm these concerns about the compatibility of pNIPAAM polymer.[71] Nevertheless, while the compatibility studies continue, the broad knowledge and experience gained from pNIPAAM studies are utilized to advance other temperature-responsive polymeric materials.[40,87]

Depending on the requirements of each application, different LCST values can be obtained by tailoring the chemical composition of the polymer. One of the early studies has shown that the LCST of pNIPAAM polymer, which is 34°C, could be increased by adding some ionic copolymers.[30] Since then, numerous studies on methods for tuning the LCST have been reported. Today, one of the most commonly used methods is copolymerization with different monomers. The LCST of the temperature-responsive polymer can be decreased by copolymerizing it with a monomer that is more hydrophobic. Similarly, the LCST can be increased by copolymerizing the temperature-responsive polymer with a monomer that is more hydrophilic.[83] For drug delivery applications the desired range of LCST is between 37°C and 42°C in order to minimize the toxicological effects due to protein denaturation.[56]

Temperature-responsive polymers find uses in other applications, such as in microfluidic devices, diagnostic devices, biocatalysts, selective separation according to size, cell cultivation, sensors, and actuators.[11,60,65] Yamato et al. have shown the usage of pNIPAAM grafts, with an LCST of 32°C, on polystyrene substrates for cell patterning applications. Accordingly, cells migrated and adhered to the areas lacking pNIPAAM below the LCST and to the grafted areas above the LCST. By systematically changing the substrate temperature and consecutively seeding, cell patterning could be achieved.[86]

3.2.2 Polymers That Are pH Responsive

Another type of stimuli-responsive polymers commonly used in tunable release studies is the pH-responsive polymers. To fully understand how these polymers can respond to pH changes, analysis of their chemical structure is crucial. Each pH-responsive polymer has acidic or basic pendant groups, which can receive or donate protons as shown in Fig. 3.9.

The polymers that contain large amounts of these side groups are ionizable and are generally defined as polyelectrolytes. Polyelectrolytes can be divided into two main groups—polyacids/polyanions (or anionic) and polybases/polycations (or cationic) polyelectrolytes—

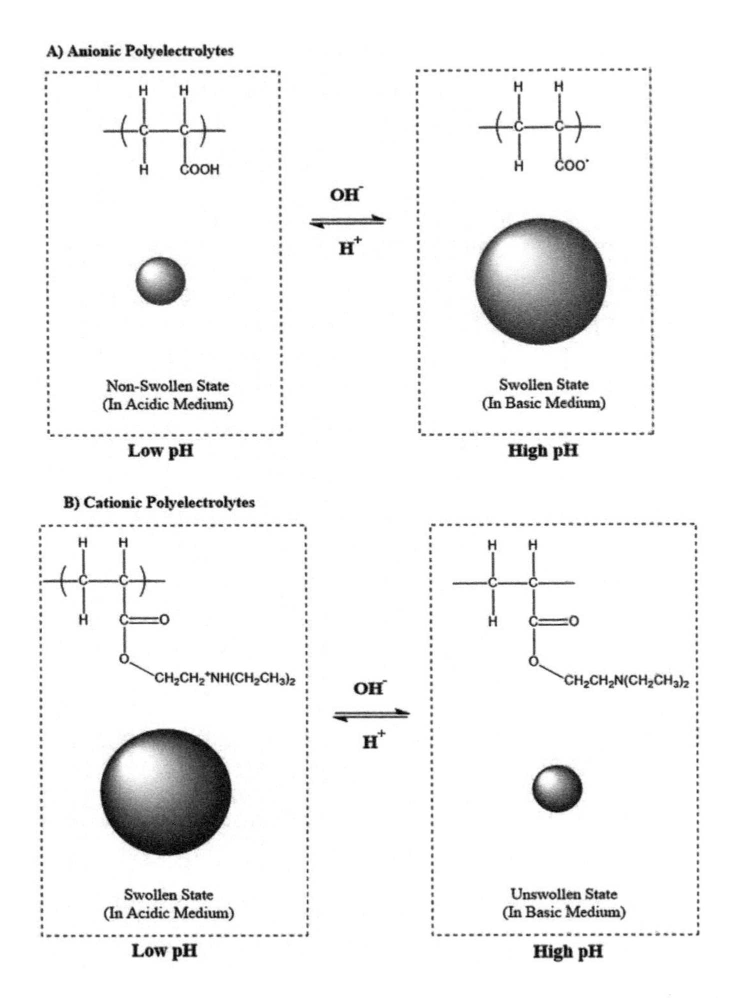

Figure 3.9 pH-sensitive ionization of anionic (a) and cationic (b) polyelectrolytes and their swelling response against pH change. Adapted from Refs. [28, 66].

depending on their pendant group. Anionic pH-responsive polymers receive protons at low pH values and donate protons under high pH conditions. pMAA and poly(acrylic acid) (pAA) are commonly used anionic pH-responsive polymers.[66] On the contrary, cationic polymers release protons at low pH values and accept protons at high pH values. Chitosan can be given as an example of cationic pH-responsive polymers.

In biomedical applications, pH is another prevalent stimulus that is used to trigger the response of polymeric systems. Different parts of the human endocrine system have different pH values, making the pH-responsive polymers suitable for use as target-specific systems. Their physical and chemical responses, like swelling/contracting and change in solubility, make these polymers especially suitable for drug delivery applications.

For pH-responsive drug delivery studies, two main approaches have been taken so far. In the first approach polymer-drug conjugates or dendrimers are formed by the utilization of pH-sensitive chemical bonds, like hydrazine. The existence of pH-labile chemical structures linking the polymer to the drug enables drug release under the low-pH environment of the extracellular and intracellular mediums of the tumor. Different structures, like polymer-drug conjugates, dendrimers, and micelles, can also be produced by means of this approach.[49]

Another approach to the design of drug delivery systems uses polymers with titratable pendant groups like carboxylic acids and amines that undergo physical or chemical dissociation following a change in the pH.[59] These approaches with facilitating different structures can be seen in Fig. 3.10.[52] The specific pH value at which a polymer with an acidic pendant group undergoes ionization is defined as pKa. An ionizable polymer with a pKa value between 3 and 10 can be a potential pH-responsive

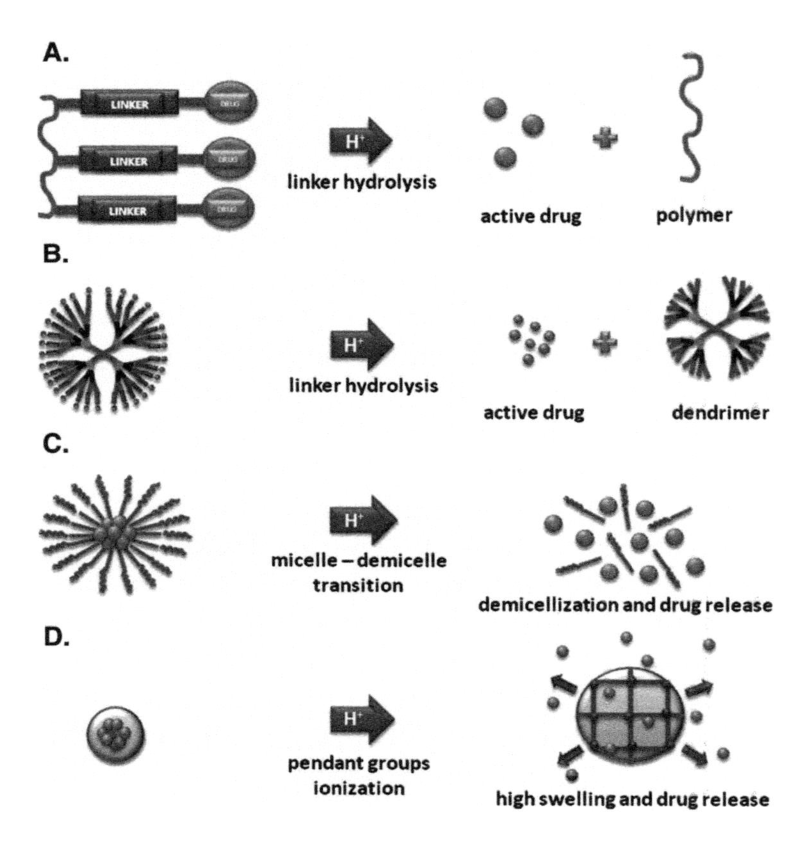

Figure 3.10 Different approaches with facilitating different structures used in drug delivery applications of pH-responsive polymers: (a) polymer-drug conjugates, (b) dendrimers, (c) micelles, and (d) nanogels or nanotubes.[52]

polymer.[73] During drug loading, the polymers are combined with cationically charged agents at pH values lower than their own pKa. Thus, they are coupled with negatively charged drugs.

As a result of the increasing repulsion forces between polymer chains, dissociation takes place and the controlled

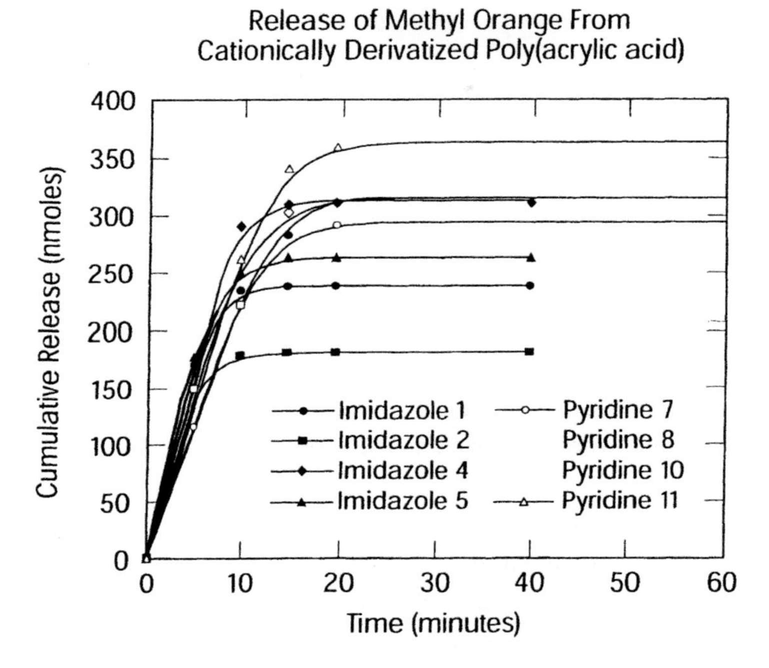

Figure 3.11 Release of methyl orange in a controlled manner from pH-responsive poly(acrylic acid) hydrogel.[62]

release of therapeutic agents occurs, as shown in Fig. 3.11, which was submitted in the patent application of Palasis.[62] In this study, insertable or implantable devices were coated with pH-responsive polymers and the negatively charged payload could be released. The coating material of pAA embodies the negatively charged therapeutic agent and includes moieties whose charge becomes positive at or below the physiological pH and becomes neutral at or above the physiological pH. This condition brings about the necessity of replacing pH-responsive moieties of the coating material with moieties that have pKa values less than the pH of 7.4. The moieties that satisfy this condition

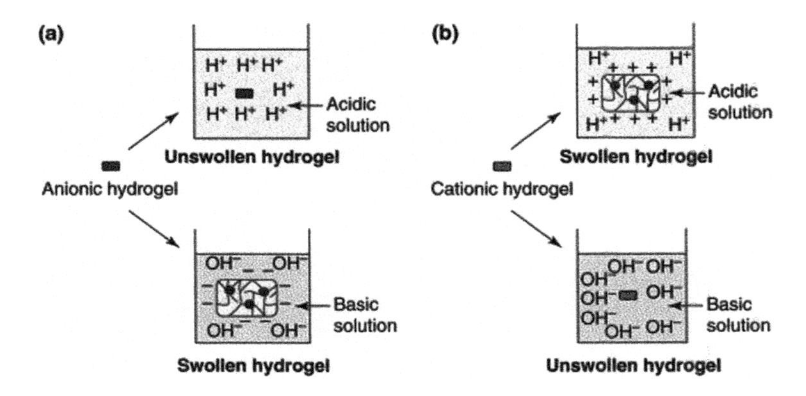

Figure 3.12 Change of the swelling degree of an anionic hydrogel and a cationic hydrogel in acidic and basic environments.[29]

and that will potentially react with the polymer include aminoethyl pyridine and aminopropyl imidazole.[62]

In Fig. 3.12, anionic and cationic hydrogels are placed in both acidic and basic environments separately. Generally, anionic pH-responsive polymers contain carboxylic acids or sulfonic acids as acidic pendant groups and their cationic counterparts contain ammonium salts. As shown in Fig. 3.12 an anionic hydrogel is protonized at a low pH and deprotonized at a high pH.[29] When exposed to a high pH, increasing electrostatic repulsion among the negatively charged groups of polymer causes swelling. This physical change triggered by pH makes the usage of the polymer as a drug delivery system possible. Exposure to different pH values not only causes the polymer to swell but also changes its solubility. In addition to the increase in the swelling ratio, hydrophobic to hydrophilic transition can also be observed for the anionic polymers at high pH values.

Most of the works related to the use of pH-responsive pMAA completed so far have utilized this polymer in the hybrid form for drug delivery applications. A study conducted by McInnes et al.[54] in which caps of a porous silicon matrix were deposited with poly[(methacrylic acid)-*co*-(ethylene dimethacrylate)], or p(MAA-*co*-EDMA), thin films can be given as an example of hybrid type of structures that are used in drug delivery applications. A porous silicon matrix was first loaded with the model drug molecule camptothecin, and then caps of the pores were deposited with the pH-sensitive copolymer p(MAA-*co*-EDMA) thin film by means of initiated CVD. After the drug was loaded onto the porous silicon matrix, a pH-responsive polymer was deposited on the pores as a cap without damaging the drug. This could be achieved by introducing the monomer in vapor phase without using any solvent and by depositing it at low temperatures.

To investigate the effect of the pH-responsive release, the drug carrying system was tested at pH 1.8 and pH 7.4. The results of the study showed that the release rates of the drug-loaded porous silicon matrix without the cap were similar at pH 1.8 and pH 7.4. However, in the case of the capped porous silicon matrix, the release rates depended strongly on the pH value, to the extent that the release of the drug was four times faster at pH 7.4 than the release rate of the same system at pH 1.8.[54] Armagan et al.[6] studied the release kinetics of p(MAA-*co*-EDMA) nanotubes using phloroglucinol (Phl) dye as the model payload at different pH conditions. The molecules were loaded onto and released from the nanotubes by increasing or decreasing the pH of the medium, which in turn caused the nanotubes to swell or collapse. The contraction of the polymer nanotubes at low pH values led to the burst release of up to 50% of the dye molecules.

The swelling behavior of the polymer in the thin film form was studied using a simple model (Eq. 3.7), applicable up to swelling ratios of 60%. The model includes the relaxation mechanism of the polymer chains due to swelling and the diffusion of the molecules from the polymer to the medium. The mathematical formulation of the model is:

$$\frac{T_t}{T_{eq}} = kt^n \tag{3.7}$$

where T_t is the change in the thickness of the film at time t, T_{eq} is the thickness measured at equilibrium, and k is a constant that depends on the material. The value of the variable n is decided upon the dominant release mechanism, as was discussed in the introduction part. Fitting the data to Eq. 3.7, the values of n were found to be 0.86 and 0.57 at pH 8 and pH 4, respectively. Due to the swelling at pH 8, the relaxation of polymer chains dominated the release mechanism, whereas at pH 4, the swelling of the polymer was negligible and the diffusion-controlled transport became dominant. After swelling proceeded for more than 60%, relaxation of the polymer became the only dominant effect on the transport of molecules through the polymer following the first-order kinetics, given by the formulation

$$\frac{T_n}{T_{eq}} = 1 - A_{\exp}\left(-k_2 t\right) \tag{3.8}$$

where A is a constant and k_2 is the rate constant for relaxation.

The release kinetics of Phl molecules from the polymer nanotubes was studied at pH 4 and pH 8, by fitting the data to the empirical formula

$$x\left(t\right) = x_f\left(1 - e^{-kt}\right) \tag{3.9}$$

where x_f is the final percentage of the dye released, $x(t)$ is the percentage of the dye at time t, and k represents the rate coefficient.[6] The values of k were found to be 0.036 min^{-1} and 0.089 min^{-1} for pH 8 and pH 4, respectively. Furthermore, at pH 4, the overall release concentration was five times higher than at pH 8. The high release rates and concentrations obtained at pH 4 were due to the contraction of the nanotubes at low pH values, leading to a stress-induced burst release.

In a study conducted by Soppimath et al. pH-responsive polyacrylamide-*graft*-(guar gum) anionic microgels were developed as containers for nifedipine (NFD) and diltiazem hydrochloride antihypertensive drugs. Release studies were performed at different pH conditions reflecting the gastric and intestinal environments.[78] Drug release from microgels of different cross-link ratios triggered by the changes in the microgel dimensions at different pH values was systematically studied, as shown in Fig. 3.13.[78] Strong dependence of the swelling extent on the cross-linking amount is shown in Fig. 3.13a. The loose structure of the network at low cross-link ratios allows accommodation of high amounts of solvent molecules, which in turn leads to higher swelling. A controlled release profile with an initial burst release can be seen in Fig. 3.13b, which was explained by the adherence of the drug during drying to the surface of the microgel. Furthermore, in Fig. 3.13c. Reversibility of pH-induced swelling is shown. The slower deswelling compared to the swelling was attributed to the effect of the deswelling on the hinderance of the proton transport. Furthermore, in order to comprehend the influence of pH on the release behavior, the release of microgels was also tested by loading a calcium channel blocker NFD and its

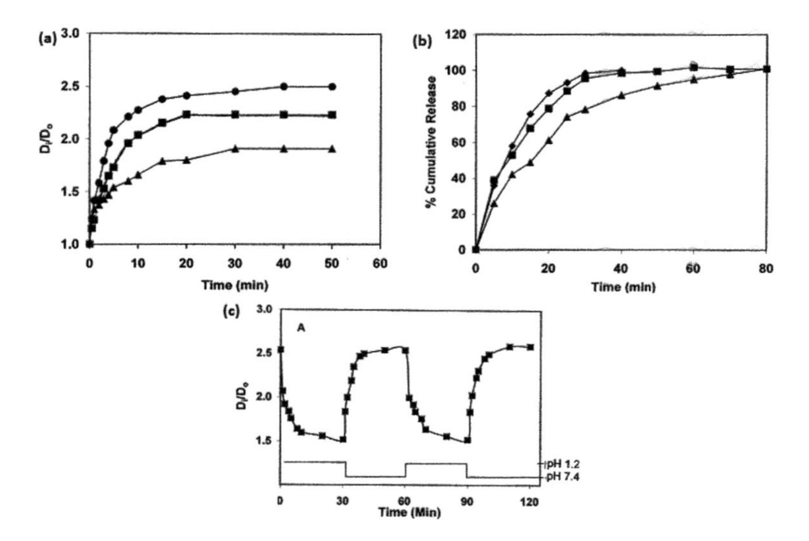

Figure 3.13 (a) Swelling behavior, (b) release profiles, and (c) pulsatile experiment of anionic hydrogels in micron size,[78] where (•) represents HGG-2.5, (■) represents HGG-5, and (▲) represents HGG-7.5.

swelling and deswelling were much faster than the release of drug-loaded microgels. Release kinetics observed were then fitted to an empirical model proposed by Robert et al.[70]:

$$\frac{D_t}{D_\infty} = kt^n \quad (3.10)$$

Here, D_t represents the microsphere diameter at time t and D_∞ represents the equilibrium diameter, k represents the rate constant, and n represents the transport mechanism type. For ionic gels, the value of n obtained was approximately 1 in a low-pH environment, indicating that case II type of transport of water was dominant. On the contrary, at pH 7.4, Fickian/non-Fickian water transport was observed.[78]

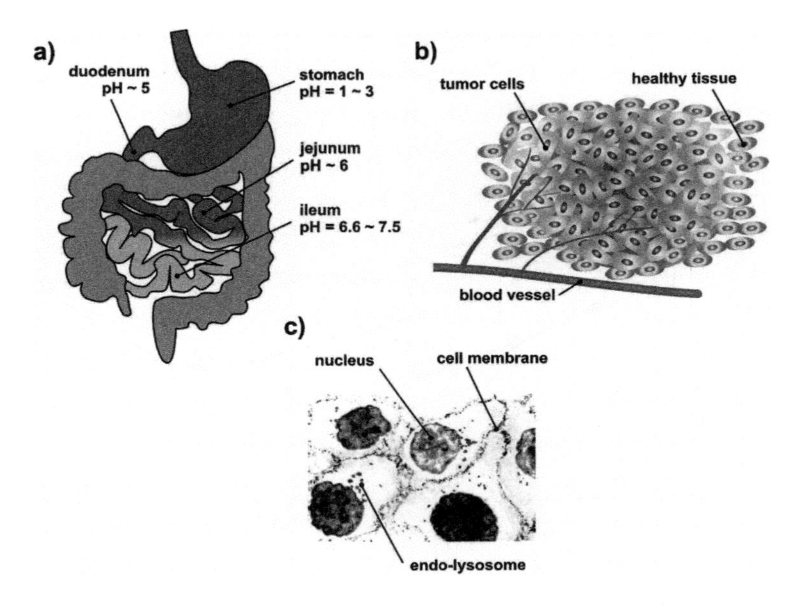

Figure 3.14 Different targeting strategies for pH-responsive polymeric drug delivery systems.[23]

Each part of the human gastrointestinal system has its own pH characteristics, and these differences in the pH enable scientists to design organ-specific drug delivery systems. It is also possible to design tissue-specific drug delivery systems that target solid tumors because of their acidic extracellular environment. Cellular-specific drug targeting, on the other hand, takes advantage of the higher acidity of endolysosomes compared to that of the cytoplasm. These different targeting systems are illustrated in Fig. 3.14.[23]

For organ-targeted drug release, mostly swelling-induced release mechanisms are employed. For instance, acrylic-based anionic-type pH-responsive pMAA nanoparticles switch to the hydrophobic, collapsed state in the

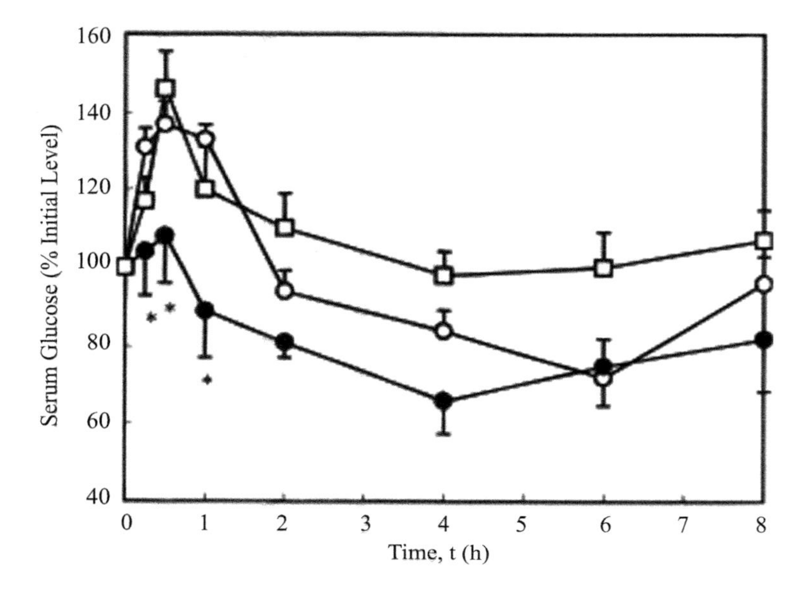

Figure 3.15 Change in the glucose response of blood after administration of pH-responsive p(MAA-*g*-EG) microspheres containing different doses of insulin.[50]

acidic environment of the stomach due to the protonation of carboxyl groups. On the other hand, the pH increase in the gastrointestinal system prompts ionization of carboxyl groups and subsequent cleavage of the hydrogen bonds, which in turn leads to swelling and drug release.[14] In the literature it is reported that swollen polymeric nanoparticles released approximately 90% of the insulin at the pH value of 7.4, whereas at pH 1.2, only 10% of the insulin is released in the collapsed state.[23]

The in-vivo studies of Lowman et al. show an increase in the glucose levels just after the drug administration (Fig. 3.15).[50] This initial increase in the glucose levels is explained by the elevated stress levels of the animals during drug administration and blood sampling. The following decrease in the glucose levels is attributed to the release of insulin in the stomach by means of structural changes in the pH-responsive polymeric gel.[50]

Detecting tumor cells is one of the main concerns of the current biomedical studies. Proliferation of tumor cells at extraordinarily high rates causes nutrient deficiencies in the tissue, resulting in high rate of glycolysis. As a result, lactic acid accumulation, which lowers the pH, begins.[40] Systems that employ pH-responsive polymers to target tumor cells in the tissue benefit from this change in the pH of the medium as a result of the aforementioned mechanisms. In the intracellular environment after endocytosis, fast endosomal acidification takes place within 2–3 min, causing the anionic polymeric structures to accept protons at low pH values. As a result, the osmotic pressure of the endosomal unit increases, initiating the drug release into the cytoplasm.[23]

Other potential applications of the pH-responsive polymeric smart material systems are personal care, membrane studies, stabilization of colloidal systems, biosensors and metal detection sensors, and encapsulation systems. To set an example, Guo et al. studied the intake conditions and amounts of Cu^{++} ions on these smart systems. Upon changes in pH, the collapse of the chain conformation with a subsequent intake of Cu^{++} ions are observed. Following the intake of cations, the whole process was optimized, serving as a logic gate for the thermometer fluorescent sensor for Cu^{++} ions.[27]

Furthermore, microfluidics is another potential application area where pH-responsive polymers are widely used. Beebe et al. designed microchannels patterned with

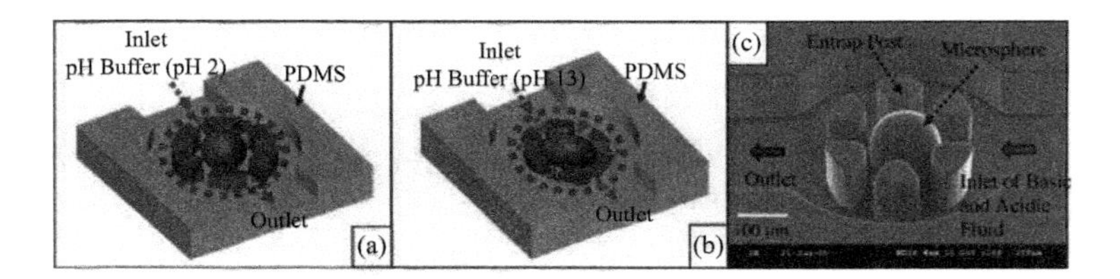

Figure 3.16 Schematic view of a microvalve in its on (a) and off (b) states upon application of alternating pH levels; (c) SEM image of the microvalve.[42]

a pH-responsive polymeric system inside. By increasing the levels of pH in the flowing solution, the polymeric hydrogel system was swollen and subsequently blocked the channels inside. On the other hand, upon decreasing the pH level of the solution, the hydrogel system collapsed and fluid could pass through the channels. This study simply highlights the exploitation of microvalve systems by using pH-responsive polymeric systems.[8] Also, Kim et al. built a system for the versatile fabrication methods employing pH-stimulated spheres in microsizes mounted in the polydimethylsiloxane-supported microfluidic valve that is shown in Fig. 3.16.[42] Following the alternating pH levels, fast and reproducible volume changes were observed and stably working microvalves were designed accordingly.[42]

3.2.3 Electroactive Polymers

Electroresponsive polymers with the characteristics of converting electrical energy to mechanical energy find applications in areas ranging from controlled drug delivery to artificial muscles.[67] Drug delivery systems that are responsive to an electric current are generally derived from polymers containing ionizable groups, called "polyelectrolytes" or "intrinsically conducting polymers." Polyelectrolytes and their pH-responsive behavior were presented earlier in detail.[7] In this section, utilization of intrinsically conductive polymers in drug delivery applications will be explained.

Conductive polymers have broad application areas because they are light in terms weight, have good electrical conductivity, can undergo a large amount of strain, have high strength, and work around room and human body temperature. For actuating them, low voltages, of 1–2 V, are sufficient and effective.[79] Biocompatibility and a stable nature make them very good candidates for numerous practical applications, and they can be produced in micro- and nanosizes easily with current technology.[79] When the conductive polymers are triggered by electrochemical reduction or oxidation, changes in the volume,[77] conductivity, or color[26,75] of the polymer may be observed. Doping of the conducting polymers can be achieved chemically or electrochemically by oxidation or reduction of the conjugated polymers. If the main chain of the conjugated polymer becomes oxidized and gives an electron, this is called "p-type electrochemical doping." On the other hand, in the n-type electrochemical doping, the main chain of the conjugated polymer is reduced and receives an electron. The doping amounts of the conjugated polymers directly affect the conductivity of the polymers.

The main drug releasing mechanisms involving conductive polymers are the mechanical force–based drug release resulting from the change in the swelling ratio of the polymeric matrix and the electrophoresis-based drug release.[67] Reversible volume change is one of main mechanisms employed in the design of mechanical drug delivery systems.

Swelling ratios of the intrinsically conducting polymers change under the influence of an external electrical field. In the case of polymer expansion, specific movements and directional forces are generated that enable them to be used as actuators with potential applications ranging from microrobotics to artificial muscles.[20]

In electrophoresis, the other main mechanism used for the delivery of the charged drug molecules, as the net electronic charge in the polymer changes as a result of an applied electric field, equivalent counterions in the medium start to move in order to keep the charge balance in the polymer. As a result, the polymer functions like an ion gate.[79] An electrostatic force is then generated on the charged drug molecules, stimulating the drug release. Meanwhile these inward and outward movements cause a change in the volume as well.[76] This volume change is directly related to the size and number of ions.[22] With the ingress of counterions the polymer expands, and with the egress of counterions the polymer contracts. This facilitates the release of the drug molecules. The polymeric matrix carrying the drug has two available redox states. One of the available redox states is suitable for binding the drug, and the other one is designed for releasing the drug. The drug release rate can easily be tuned by adjusting properties of the polymer like conductivity, volume, and charge. Electrostatic release mechanisms of an anionic drug and a cationic drug are illustrated in Fig. 3.17.[81]

A recent approach in developing new drug delivery systems focuses on utilizing conductive polymer nanotubes with large surface areas. In a representative study, Abidian et al. investigated controlled release characteristics of dexamethasone-loaded poly(3,4-ethylenedioxythiophene),

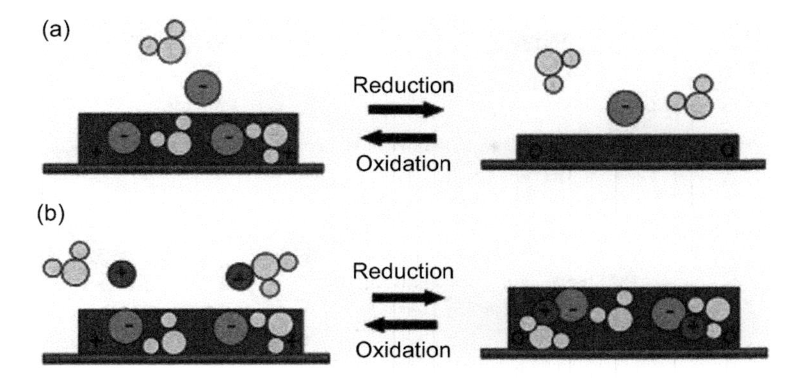

Figure 3.17 Electrostatic release mechanisms of (a) an anionic drug and (b) a cationic drug.[81]

or PEDOT, conductive polymeric nanotubes, which were prepared by coating on the surfaces of sacrificial electrospun PLGA nanofibers.[1] The increased surface area of the textured exterior surface of the nanotubes enhanced the transfer between ions and electrons. For release actuation, a 1 V positive voltage was applied to the PEDOT nanotubes at specific times.

In response to the positive bias applied on the conductive polymer nanotubes, electrons inside the solution were attracted to the polymer chains to maintain the charge neutrality of the system; whereas ions with negative charges were repelled into the solution. Contraction of the PEDOT nanotubes induced the buildup of stress, and the hydrodynamical force acting on the PLGA triggered the drug release. Besides the slow degradation of PLGA, swelling of the polymer due to changes in the electric field influences the drug release kinetics. Approximately 75% of the loaded drug was released in the first seven days, and after fifty-four days had passed, the remaining 25% of the

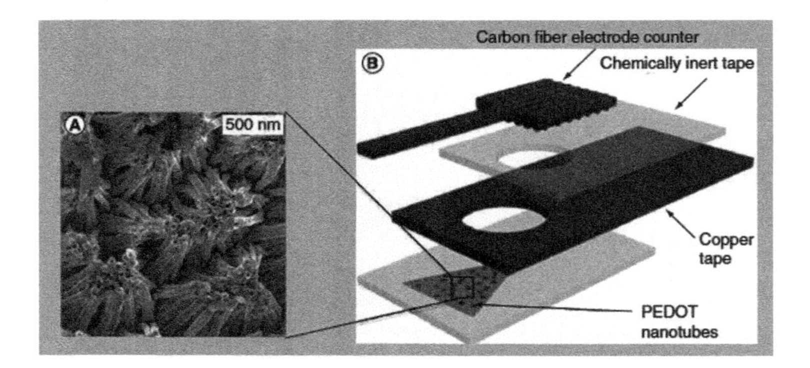

Figure 3.18 (A) Scanning electron microscope micrograph of PEDOT nanotubes and (B) PEDOT nanotubes including a transdermal patch for ex vivo studies.[58]

dexamethasone was still being released, which indicates that the PEDOT shield delayed the release substantially.[1]

Release of hydrophilic drugs in transdermal applications is quite challenging due to the lipophilic nature of the transdermal medium. In a study by Nguyen et al., to address this challenge, a transdermal patch with PEDOT nanotubes for ex vivo transdermal experiments was designed to investigate the release kinetics of two model drugs and insulin. The schematic of this patch is depicted in Fig. 3.18.[58]

In addition to the insulin, PEDOT nanotubes were also loaded with the model anionic drugs amaranth and Cy5.5. During the release experiments, for each payload system, besides the negative voltage applied to initiate the release of anionic drugs, positive and zero voltages were also applied to study the preventative effect and the control conditions, respectively. Their release constants were determined separately, where k_r represents the linear release rate constant, k_c represents the control release rate constant, and k_p represents the preventative release rate constant in Fig. 3.19.[58] Since amaranth has a charge of −3 and Cy5.5 has a charge of −1 charge (at pH 7.4) and Cy5.5 is less hydrophilic compared to amaranth, higher voltage potentials were applied for the Cy 5.5 drug in order to stimulate the release. Furthermore, to make a comparison and comprehend the effects of nanotubes, PEDOT thin films were also used in the release experiments. The data collected from these measurements are as shown in Fig. 3.19.[58]

In the release experiments performed for more than 24 hours, the quantity of the insulin released was observed to be significantly less than that of the other two drugs, though the release kinetics of all the drugs were similar. For each system, controlled release characteristics, which include an initial burst release followed by a long-term steady state release, were observed. It was observed that when an external electrical potential is applied, the released quantity was comparable to the amounts released in the case when the drug was administered with an injection. This finding relates the amount of drug molecules released directly to the number of pulses applied.

When the release kinetics from these samples were compared to the kinetics of control samples to which no electrical potential was applied, it was observed that the usage of electrical stimulation induced three times faster release rates and release amounts four times larger than the amounts obtained in the absence of stimulation. Besides, the amount of drug released from the PEDOT nanotubes was three times larger than the amounts released from PEDOT films. Electroresponsive polymeric systems for drug delivery applications are not fully commercialized yet. However, the development of drug delivery systems

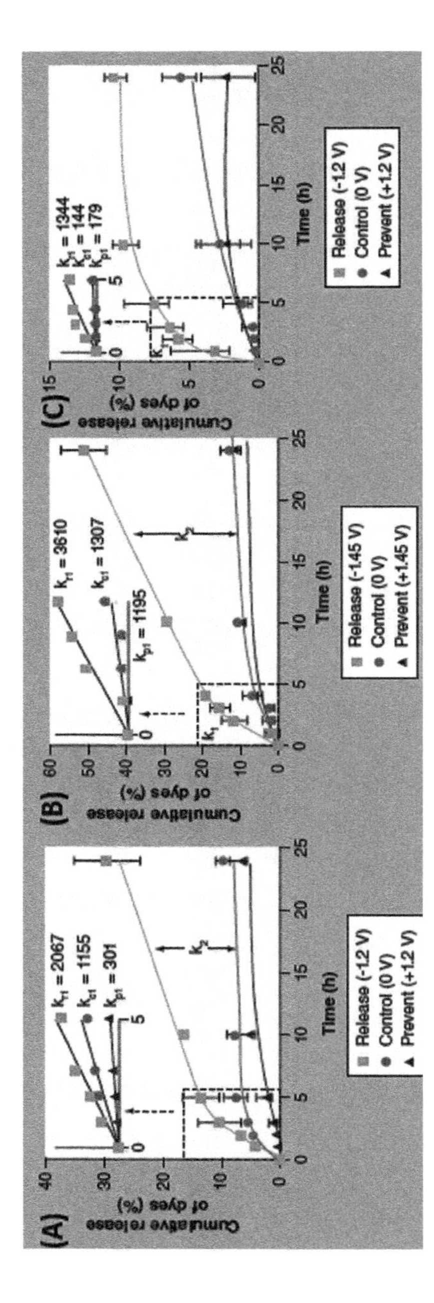

Figure 3.19 Drug release kinetics observed in porcine transdermal on using (A) amaranth, (B) Cy5.5, and (C) insulin.[58]

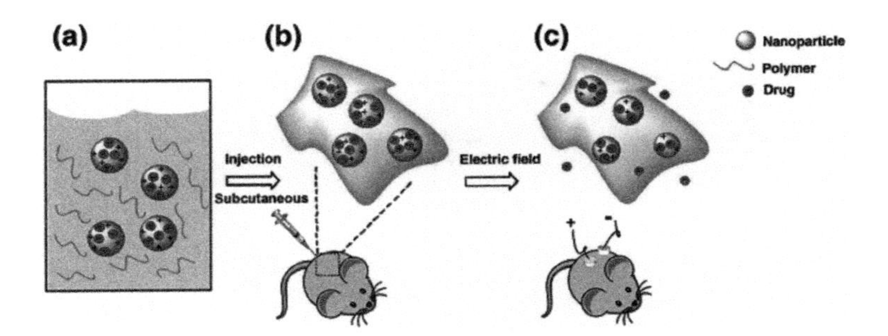

Figure 3.20 External stimulation of an injectable polymer-nanoparticle-coupled electric field responsive system.[25]

that couple these responsive polymers with micro- or nano-sized electronic systems, which can be triggered by an external source as shown in Fig. 3.20, would significantly impact the field.[25] In this system, conductive polymer nanoparticles to be used as drug reservoirs are first loaded with a therapeutic agent. After the drug carrying nanoparticles are transferred to the mouse via a subcutaneous injection, controlled release is triggered by an external electric field.[81] For example, it would be possible in clinical applications to control an implantable system for in vivo drug delivery via wireless signals.[17]

3.2.4 Light-Responsive Polymers

Light responsive polymers are widely used in controlled release studies where the release of the payload is activated by light. These polymers undergo structural changes when exposed to electromagnetic radiation. In applications, mostly electromagnetic radiation in the ultraviolet (UV) and visible range is used to trigger the response and the responsive polymers are generally categorized as visible light–sensitive and UV-sensitive polymers.

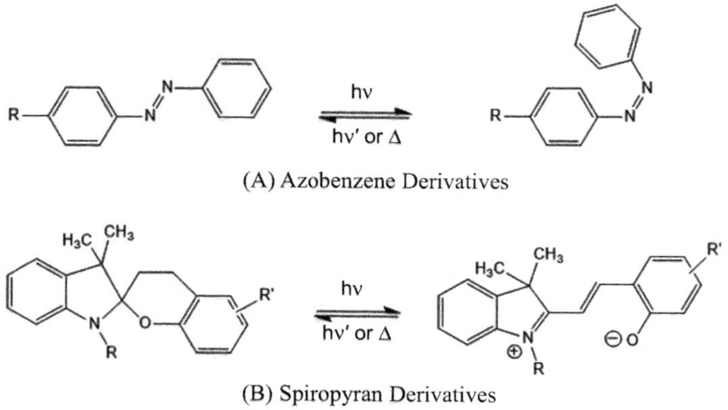

(A) Azobenzene Derivatives

(B) Spiropyran Derivatives

Figure 3.21 Light-triggered reversible transformations of (A) azobenzene and (B) spiropyran derivatives.[93]

Polymers that include chromophore moieties like trisodium salt of copper chlorophyllin exhibit conformational changes under visible light and thus belong to the group of visible light–sensitive polymers.[67] One of the most widely studied chromophores is azobenzene. The chemical and structural changes azobenzene undergoes are illustrated in Fig. 3.21, and the mechanisms behind the changes are well studied in the literature.[53,57] Light exposure causes the *trans-cis* isomerization, and as a result, the concentration of chromophores decreases along the polarization direction. Polar domains are reoriented by the chromophore motion at the domain scale, and finally at the mass level macroscopic motions of the polymer can be observed and it can be applied for surface patterning. As a result of this *cis*-to-*trans* isomerization, the shape and the electronic structure of azobenzene change.[57]

In addition to azobenzene, derivatives of spiropyran can also demonstrate visible light–sensitive behavior. Although spiropyran groups are comparatively nonpolar, irradiation with proper wavelength causes formation of zwitterionic isomers, which have a considerably higher dipole moment.[93]

UV-sensitive polymers, on the other hand, contain leuco-derivative moieties. For instance, the 4-dimethylamino phenylmethyl leucocyanide molecule undergoes phase transition when exposed to UV radiation. Under UV light stimulation, dissociation of molecules leads to the formation of cyanide ions at a constant temperature and increases the osmotic pressure within the polymeric hydrogel, causing swelling. In the case of triphenylmethane-leuco, the molecule disintegrates into ion pairs and triphenylmethyl cations form upon exposure to UV radiation, leading to an increase in the osmotic pressure, which causes the hydrogel containing these moieties to swell. The presence of cyanide ions is thought to be the cause of the osmotic pressure increase. When the UV radiation is turned off, the hydrogel collapses, confirming the reversible nature of the response mechanism.[67] Visible light–sensitive polymers are preferred to the UV-sensitive polymers for controlled release applications due to availability, biocompatibility, and safety of the visible light.

Two main approaches are taken when light-sensitive polymers are used as nanocarriers for drug delivery applications, as depicted in Fig. 3.22.[18] In one approach (Fig. 3.22a) core-shell structures are designed with core parts containing drug-loaded bioactive molecules. The release of the entrapped drug molecules occurs when the outer shell undergoes cleavage- or degradation-type structural changes or variations in the charges of the functional groups. In the other approach, which involves conjugation of the drug with the nanocarrier (Fig. 3.22b), the drug

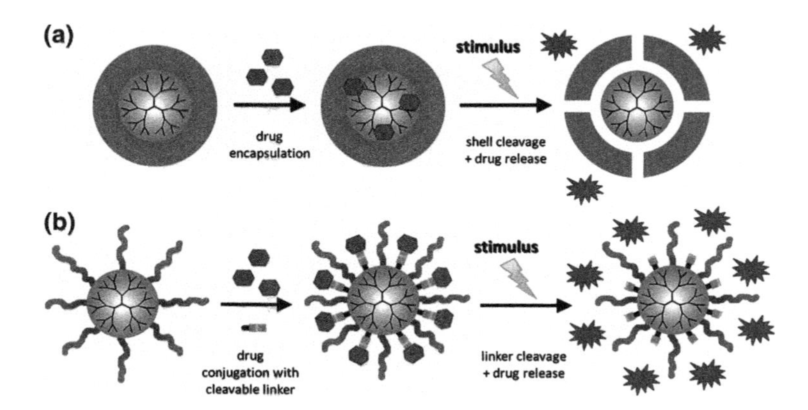

Figure 3.22 Two main approaches to be used as nanocarriers for drug delivery applications.[18]

release is activated by the cleavage of linker molecules between the bioactive molecules and the carrier.[18]

Benefiting from the molecular switch properties of azobenzene, drug delivery systems containing cleavable linkers or protective shell structures that can be activated by light could be developed. Spatial and temporal control of the release by using a stimulus of light is possible in these systems.[2] While some light-sensitive drug delivery systems are for single use due to the irreversible conformational changes and the burst release of the entire drug occurring at a single stimulation, other delivery systems can undergo a reversible change and when exposed to light and dark cycles, they can release drugs in a pulsatile manner, making them potential multiswitchable therapeutic agent carriers.[4]

Xiao et al. reported a promising approach for cancer treatment involving dual stimuli–sensitive capsules in microsizes produced using the layer-by-layer technique. Shell layers were prepared from polymers of

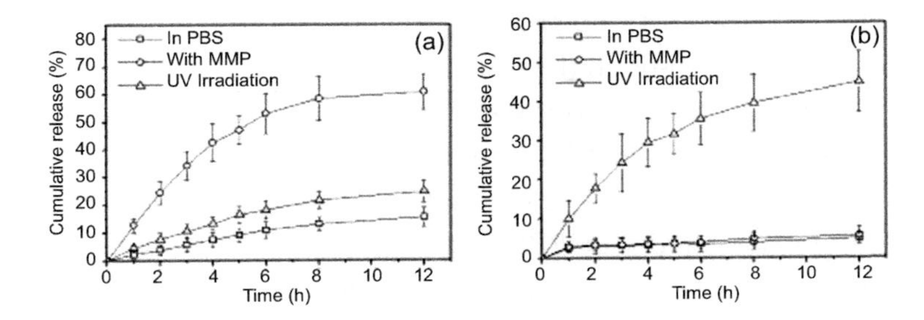

Figure 3.23 Release profiles of (a) a macromolecular drug dextran5000–fluorescein isothiocyanate (Dex5000-FITC) and (b) a small molecule a-CD–rhodamine B (a-CD-RhB) under varied stimuli.[85]

poly[(acrylic acid)-*graft*-azobenzene-*graft*-proline-leucine -glycine-valine-arginine-adamantane] and poly[(aspartic acid)-*graft-b*-cyclodextrin]. Utilizing the *cis*-to-*trans* phase transition of azobenzene, the release kinetics of a macromolecular drug dextran5000–fluorescein isothiocyanate (Dex5000-FITC) and a small molecule a-CD–rhodamine B (a-CD-RhB) were investigated under UV irradiation. The results are as shown in Fig. 3.23.[85]

The release experiments were performed in metalloproteinase (MMP) solution and phosphate buffer solution (PBS) separately, and UV irradiation was applied to stimulate the release from the multilayer microcapsules. During the release studies of the macromolecular drug, it was observed that only 15% of the drug was released in the first 12 hours from the microcapsules in the PBS solution without a light stimulus (Fig. 3.23a). Similar release kinetics were observed for the macromolecular drug in the presence of UV irradiation. On the other hand, with MMP, which is a protease enzyme used in tumor targeting systems as an internal stimulus, the release percentage of the drug increased to up to 60%, indicating the considerable influence of MMP on the release.

Studies on the release kinetics of small molecules have shown that the UV irradiation significantly increased the release rate and the amount of molecules released. While the effects of PBS and MMP on the release were negligible, release percentages of up to 40% were observed under UV exposure. The study demonstrated that different drugs loaded in the same dual-responsive system could be released under different stimuli, like MMP or UV.[85] UV irradiation cannot be used for extended time periods in biorelated applications. To overcome this limitation, metal nanoparticles of anisotropic geometries, which exhibit enhanced near-infrared radiation absorption, can be incorporated into the light-sensitive polymer structures to facilitate the release process.[41] Furthermore, dual- and multiresponsive polymers can be developed using two or more polymers that respond to different stimuli. The majority of these kinds of studies involve a combination of light- and thermoresponsive polymers.

In one of the earliest studies conducted by Kungwatchakun and Irie[45] copolymerization of *N*-(4-phenylazophenyl) acrylamide and *N*-isopropyl acrylamide (NIPAAM) was carried out to control the LCST of pNIPAAM via light stimuli. Introducing azobenzene molecules into the NIPAAM, LCST could be shifted from 21°C to 27°C. Shifting of the phase transition temperature was explained by the dipole moment change, which is the result of *cis*-to-*trans* isomerization that azobenzene molecules undergo.

The light-responsive polymers have aroused interest and have been studied for a wide range of potential applications. For example, considering that tissue parts are transparent to near-infrared radiation, they can be successfully used in biomedical research. One of the disadvantages of the light-responsive polymers as controlled release platforms is the inconsistent response due to the leaching of the noncovalently bounded chromophores during the process of expansion or contraction of the whole system, resulting in the weak or slow response of the polymer system.[67] Furthermore, in order to avoid an early and unexpected release or premature activation leading to toxic side effects the patient must be kept in a dark environment before the treatment.[18] This inconvenience on the patient side is another disadvantage of using light-responsive polymers in in vivo delivery applications.

3.3 Summary

The last decade has seen great advances in the field of stimuli-responsive, smart polymers. New polymers that are responsive to different stimuli, such as temperature, pH, and light, are designed in the laboratories to satisfy the requirements of specific applications. The ability to tune the response of a polymer to a stimulus by tailoring the chemical or structural composition of the polymer significantly widens the application areas of these smart materials. From medicine to electronics, these polymers are slowly replacing the conventional materials.

One of the major applications in which these polymers have been widely utilized is controlled release, which is critical in the fields of drug delivery, cosmetics, food, and agriculture. Developments in the field of smart polymers had the highest impact on drug delivery. Depending on the specifics of the drug delivery application, such as the type of drug or target tissue, certain characteristics of the delivery platform can be tailored by the integration of the stimuli-responsive polymers into these delivery systems. The characteristics that need to be tailored include release rates and release amounts of the drugs, the duration of the release, and the target location. The degree of the physicochemical change that the polymer undergoes when triggered by a stimulus can be controlled by tuning the chemical composition of the polymer. Control of the physicochemical response of the polymer enables the researchers to directly address the needs of the delivery application. Current state-of-the-art delivery systems involve polymers with well-controlled release kinetics, which are responsive to multiple stimuli for more efficient activation.

Despite all these advances in the drug delivery field with the usage of stimuli-responsive polymers, new challenges arise each day with the development of new treatment methods. These challenges can be met by further research on stimuli-responsive polymers, which prove to be one of the most effective means of delivery.

References

1. Abidian, M. R., Kim, D. H. and Martin, D. C. (2006). Conducting-polymer nanotubes for controlled drug release, *Adv. Mater.*, **18**(4), pp. 405–409.
2. Aghabegi Moghanjoughi, A., Khoshnevis, D. and Zarrabi, A. (2016). A concise review on smart polymers for controlled drug release, *Drug Deliv. Transl. Res.*, **6**(3), pp. 333–340.
3. Al-Tahami, K. and Singh, J. (2007). Smart polymer based delivery systems for peptides and proteins, *Recent Pat. Drug Deliv. Formul.*, **1**(1), pp. 65–71.

4. Alvarez-Lorenzo, C., Bromberg, L. and Concheiro, A. (2009). Light-sensitive intelligent drug delivery systems, *Photochem. Photobiol.*, **85**(4), pp. 848–860.

5. Armagan, E. and Ozaydin Ince, G. (2015). Coaxial nanotubes of stimuli responsive polymers with tunable release kinetics, *Soft Matter*, **11**(41), pp. 8069–8075.

6. Armagan, E., Qureshi, P. and Ince, G. O. (2015). Functional nanotubes for triggered release of molecules, *Nanosci. Nanotechnol. Lett.*, **7**(1), pp. 79–83.

7. Bajpai, A. K., Shukla, S. K., Bhanu, S. and Kankane, S. (2008). Responsive polymers in controlled drug delivery, *Prog. Polym. Sci.*, **33**(11), pp. 1088–1118.

8. Beebe, D. J., Moore, J. S., Bauer, J. M., Yu, Q., Liu, R. H., Devadoss, C. and Jo, B.-H. (2000). Functional hydrogel structures for autonomous flow control inside microfluidic channels, *Nature*, **404**(6778), pp. 588–590.

9. Bravo, S. A., Lamas, M. C. and Salomón, C. J. (2002). In-vitro studies of diclofenac sodium controlled-release from biopolymeric hydrophilic matrices, *J. Pharm. Pharm. Sci.*, **5**(3), pp. 213–219.

10. Cai, K., Jiang, F., Luo, Z. and Chen, X. (2010). Temperature-responsive controlled drug delivery system based on titanium nanotubes, *Adv. Eng. Mater.*, **12**(9), pp. B565–B570.

11. Chaterji, S., Kwon, I. K. and Park, K. (2007). Smart polymeric gels: Redefining the limits of biomedical devices, *Prog. Polym. Sci.*, **32**(8–9), pp. 1083–1122.

12. Chen, G., Chen, R., Zou, C., Yang, D. and Chen, Z.-S. (2014). Fragmented polymer nanotubes from sonication-induced scission with a thermo-responsive gating system for anti-cancer drug delivery, *J. Mater. Chem. B*, **2**(10), pp. 1327–1334.

13. Choi, S., Baudys, M. and Kim, S. W. (2004). Control of blood glucose by novel GLP-1 delivery using biodegradable triblock copolymer of PLGA-PEG-PLGA in type 2 diabetic rats, *Pharm. Res.*, **21**(5), pp. 827–831.

14. Colombo, P., Sonvico, F., Colombo, G. and Bettini, R. (2009). Novel platforms for oral drug delivery, *Pharm. Res.*, **26**(3), pp. 601–611.

15. Costa, P. and Sousa Lobo, J. M. (2001). Modeling and comparison of dissolution profiles, *Eur. J. Pharm. Sci.*, **13**(2), pp. 123–133.

16. Crespy, D. and Rossi, R. M. (2007). Temperature-responsive polymers with LCST in the physiological range and their applications in textiles, *Polym. Int.*, **56**(12), pp. 1461–1468.

17. Dissanayake, T. D., Hu, A. P., Malpas, S., Bennet, L., Taberner, A., Booth, L. and Budgett, D. (2009). Experimental study of a TET system for implantable biomedical devices, *IEEE Trans. Biomed. Circuits Syst.*, **3**(6), pp. 370–378.

18. Fleige, E., Quadir, M. A. and Haag, R. (2012). Stimuli-responsive polymeric nanocarriers for the controlled transport of active compounds: concepts and applications, *Adv. Drug Deliv. Rev.*, **64**(9), pp. 866–884.

19. Folkman, J. (1990). How the field of controlled-release technology began, and its central role in the development of angiogenesis research, *Biomaterials*, **11**(9), pp. 615–618.

20. Fuchiwaki, M., Martinez, J. G. and Otero, T. F. (2015). Polypyrrole asymmetric bilayer artificial muscle: driven reactions, cooperative actuation, and osmotic effects, *Adv. Funct. Mater.*, **25**(10), pp. 1535–1541.

21. Gandhi, A., Paul, A., Sen, S. O. and Sen, K. K. (2015). Studies on thermoresponsive polymers: Phase behaviour, drug delivery and biomedical applications, *Asian J. Pharm. Sci.*, **10**(2), pp. 99–107.

22. Gandhi, M. R., Murray, P., Spinks, G. M. and Wallace, G. G. (1995). Mechanism of electromechanical actuation in polypyrrole, *Synth. Met.*, **73**(3), pp. 247–256.

23. Gao, W., Chan, J. M. and Farokhzad, O. C. (2010). pH-responsive nanoparticles for drug delivery, *Mol. Pharmaceutics*, **7**(6), pp. 1913–1920.

24. Gao, Y., Zhou, Y. and Yan, D. (2009). Temperature-sensitive and highly water-soluble titanate nanotubes, *Polymer*, **50**(12), pp. 2572–2577.

25. Ge, J., Neofytou, E., Cahill, T. J., Beygui, R. E. and Zare, R. N. (2012). Drug release from electric field responsive nanoparticles, *ACS Nano*, **6**(1), pp. 227–233.

26. Groenendaal, L., Jonas, F., Freitag, D., Pielartzik, H. and Reynolds, J. R. (2000). Poly (3, 4-ethylenedioxythiophene) and its derivatives: past, present, and future, *Adv. Mater.*, **12**(7), pp. 481–494.

27. Guo, Z., Zhu, W., Xiong, Y. and Tian, H. (2009). Multiple logic fluorescent thermometer system based on N-isopropylmethacrylamide copolymer bearing dicyanomethylene-4H-pyran moiety, *Macromolecules*, **42**(5), pp. 1448–1453.

28. Gupta, K. and Gangadharappa, H. (2011). pH sensitive drug delivery systems: a review, *Am. J. Drug Discovery Dev.*, **1**(1), pp. 24–48.

29. Gupta, P., Vermani, K. and Garg, S. (2002). Hydrogels: from controlled release to pH-responsive drug delivery, *Drug Discovery Today*, **7**(10), pp. 569–579.

30. Hirotsu, S., Hirokawa, Y. and Tanaka, T. (1987). Volume-phase transitions of ionized N-isopropylacrylamide gels, *J. Chem. Phys.*, **87**(2), pp. 1392–1395.

31. Hixson, A. W. and Crowell, J. H. (1931). Dependence of reaction velocity upon surface and agitation, *Ind. Eng. Chem.*, **23**(8), pp. 923–931.

32. Hoffman, A. S. (2000). Bioconjugates of intelligent polymers and recognition proteins for use in diagnostics and affinity separations, *Clin. Chem.*, **46**(9), pp. 1478–1486.

33. Hrubý, M., Filippov, S. K. and Štěpánek, P. (2015). Smart polymers in drug delivery systems on crossroads: Which way deserves following? *Eur. Polym. J.*, **65**: 82–97.

34. Huang, X. and Brazel, C. S. (2001). On the importance and mechanisms of burst release in matrix-controlled drug delivery systems, *J. Control. Release*, **73**(2–3), pp. 121–136.

35. Hurwitz, M. and Stauffer, P. (2014). Hyperthermia, radiation and chemotherapy: the role of heat in multidisciplinary cancer care, *Semin. Oncol.*, **41**(6), pp. 714–729.

36. Jaafar, M., Razak, K. A., Ariff, Z. M., Zabidi, H., Hilmi, B., Hamid, Z. A. A., Akil, H. M. and Yahaya, B. H. (2016). 5th international conference on recent advances in materials, minerals and environment (RAMM) & 2nd international postgraduate conference on materials, mineral and polymer (MAMIP). The characteristics of the smart polymeras temperature or pH-responsive hydrogel, *Procedia Chem.*, **19**, pp. 406–409.

37. James, J. R., Gao, Y., Soon, V. C., Topper, S. M., Babsky, A. and Bansal, N. (2010). Controlled radio-frequency hyperthermia using an MR scanner and simultaneous monitoring of temperature and therapy response by 1H, 23Na and 31P magnetic resonance spectroscopy in subcutaneously implanted 9L-gliosarcoma, *Int. J. Hyperther.*, **26**(1), pp. 79–90.

38. Janssen, M., Mihov, G., Welting, T., Thies, J. and Emans, P. (2014). Drugs and polymers for delivery systems in OA joints: clinical needs and opportunities, *Polymers*, **6**(3), p. 799.

39. Jones, E., Thrall, D., Dewhirst, M. W. and Vujaskovic, Z. (2006). Prospective thermal dosimetry: the key to hyperthermia's future, *Int. J. Hyperther.*, **22**(3), pp. 247–253.

40. Kamaly, N., Yameen, B., Wu, J. and Farokhzad, O. C. (2016). Degradable controlled-release polymers and polymeric nanoparticles: mechanisms of controlling drug release, *Chem. Rev.*, **116**(4), pp. 2602–2663.

41. Kang, H., Trondoli, A. C., Zhu, G., Chen, Y., Chang, Y.-J., Liu, H., Huang, Y.-F., Zhang, X. and Tan, W. (2011). Near-infrared light-responsive core–shell nanogels for targeted drug delivery, *ACS Nano*, **5**(6), pp. 5094–5099.

42. Kim, D., Kim, S., Park, J., Baek, J., Kim, S., Sun, K., Lee, T. and Lee, S. (2007). Hydrodynamic fabrication and characterization of a pH-responsive microscale spherical actuating element, *Sens. Actuators, A*, **134**(2), pp. 321–328.

43. Kochba, M., Gambash, S. and Avnimelech, Y. (1990). Studies on slow release fertilizers: 1. effects of temperature, soil moisture, and water vapor pressure, *Soil Sci.*, **149**(6), pp. 339–343.

44. Korsmeyer, R. W., Gurny, R., Doelker, E., Buri, P. and Peppas, N. A. (1983). Mechanisms of solute release from porous hydrophilic polymers, *Int. J. Pharm.*, **15**(1), pp. 25–35.

45. Kungwatchakun, D. and Irie, M. (1988). Photoresponsive polymers. Photocontrol of the phase separation temperature of aqueous solutions of poly-[N-isopropylacrylamide-co-N-(4-phenylazophenyl)acrylamide], *Makromol. Chem. Rapid Commun.*, **9**(4), pp. 243–246.

46. Lendlein, A. and Shastri, V. P. (2010). Stimuli-sensitive polymers, *Adv. Mater.*, **22**(31), pp. 3344–3347.

47. Li, Y. Y., Zhang, X. Z., Zhu, J. L., Cheng, H., Cheng, S. X. and Zhuo, R. X. (2007). Self-assembled, thermoresponsive micelles based on triblock PMMA-b-PNIPAAm-b-PMMA copolymer for drug delivery, *Nanotechnology*, **18**(21), p. 215605.

48. Li, Z., Tang, Y.-H., Li, X. and Karniadakis, G. E. (2015). Mesoscale modeling of phase transition dynamics of thermoresponsive polymers, *Chem. Commun.*, **51**(55), pp. 11038–11040.

49. Liu, Y., Wang, W., Yang, J., Zhou, C. and Sun, J. (2013). pH-sensitive polymeric micelles triggered drug release for extracellular and intracellular drug targeting delivery, *Asian J. Pharm. Sci.*, **8**(3), pp. 159–167.

50. Lowman, A. M., Morishita, M., Kajita, M., Nagai, T. and Peppas, N. A. (1999). Oral delivery of insulin using pH-responsive complexation gels, *J. Pharm. Sci.*, **88**(9), pp. 933–937.

51. Mahajan, A. and Aggarwal, G. (2011). Smart polymers: innovations in novel drug delivery, *Int. J. Drug Dev. Res.*, **3**(3), pp. 16-30.

52. Manchun, S., Dass, C. R. and Sriamornsak, P. (2012). Targeted therapy for cancer using pH-responsive nanocarrier systems, *Life Sci.*, **90**(11–12), pp. 381–387.

53. Matějka, L., Ilavský, M., Dušek, K. and Wichterle, O. (1981). Photomechanical effects in crosslinked photochromic polymers, *Polymer*, **22**(11), pp. 1511–1515.

54. McInnes, S. J. P., Szili, E. J., Al-Bataineh, S. A., Xu, J., Alf, M. E., Gleason, K. K., Short, R. D. and Voelcker, N. H. (2012). Combination of iCVD and porous silicon for the development of a controlled drug delivery system, *ACS Appl. Mater. Interfaces*, **4**(7), pp. 3566–3574.

55. Mittal, V., Matsko, N. B., Butté, A. and Morbidelli, M. (2007). Synthesis of temperature responsive polymer brushes from polystyrene latex particles functionalized with ATRP initiator, *Eur. Polym. J.*, **43**(12), pp. 4868–4881.

56. Mura, S., Nicolas, J. and Couvreur, P. (2013). Stimuli-responsive nanocarriers for drug delivery, *Nat. Mater.*, **12**(11), pp. 991–1003.

57. Natansohn, A. and Rochon, P. (2002). Photoinduced motions in azo-containing polymers, *Chem. Rev.*, **102**(11), pp. 4139–4176.

58. Nguyen, T. M., Lee, S. and Lee, S. B. (2014). Conductive polymer nanotube patch for fast and controlled ex vivo transdermal drug delivery, *Nanomedicine*, **9**(15), pp. 2263–2272.

59. O'Neill, B. E. and Li, K. C. P. (2008). Augmentation of targeted delivery with pulsed high intensity focused ultrasound, *Int. J. Hyperther.*, **24**(6), pp. 506–520.

60. Okano, T., Bae, Y. H., Jacobs, H. and Kim, S. W. (1990). Thermally on-off switching polymers for drug permeation and release, *J. Control. Release*, **11**(1), pp. 255–265.

61. Ozaydin-Ince, G., Gleason, K. K. and Demirel, M. C. (2011). A stimuli-responsive coaxial nanofilm for burst release, *Soft Matter*, **7**(2), pp. 638–643.

62. Palasis, M. (2003). Implantable or insertable therapeutic agent delivery device, Google Patents.

63. Peppas, N. A. (1985). Analysis of Fickian and non-Fickian drug release from polymers, *Pharm. Acta Helv.*, **60**(4), pp. 110–111.

64. Peppas, N. A., Bures, P., Leobandung, W. and Ichikawa, H. (2000). Hydrogels in pharmaceutical formulations, *Eur. J. Pharm. Biopharm.*, **50**(1), pp. 27–46.

65. Peppas, N. A., Hilt, J. Z., Khademhosseini, A. and Langer, R. (2006). Hydrogels in biology and medicine: from molecular principles to bionanotechnology, *Adv. Mater.*, **18**(11), pp. 1345–1360.

66. Priya, B., Viness, P., Yahya, E. C. and Lisa, C. d. T. (2009). Stimuli-responsive polymers and their applications in drug delivery, *Biomed. Mater.*, **4**(2), p. 022001.

67. Priya James, H., John, R., Alex, A. and Anoop, K. R. (2014). Smart polymers for the controlled delivery of drugs – a concise overview, *Acta Pharm. Sin. B*, **4**(2), pp. 120–127.

68. Qiu, H., Wan, M., Matthews, B. and Dai, L. (2001). Conducting polyaniline nanotubes by template-free polymerization, *Macromolecules*, **34**(4), pp. 675–677.

69. Qiu, Y. and Park, K. (2001). Environment-sensitive hydrogels for drug delivery, *Adv. Drug Deliv. Rev.*, **53**(3), pp. 321–339.

70. Robert, C. C. R., Buri, P. A. and Peppas, N. A. (1985). Effect of degree of crosslinking on water transport in polymer microparticles, *J. Appl. Polym. Sci.*, **30**(1), pp. 301–306.

71. Shakya, A. K., Kumar, A. and Nandakumar, K. S. (2011). Adjuvant properties of a biocompatible thermo-responsive polymer of N-isopropylacrylamide in autoimmunity and arthritis, *J. R. Soc. Interface*, **8**(65), pp. 1748–1759.

72. Shen, Z., Wei, W., Zhao, Y., Ma, G., Dobashi, T., Maki, Y., Su, Z. and Wan, J. (2008). Thermosensitive polymer-conjugated albumin nanospheres as thermal targeting anti-cancer drug carrier, *Eur. J. Pharm. Sci.* **35**(4), pp. 271–282.

73. Siegel, R. A. (1993). Hydrophobic weak polyelectrolyte gels: studies of swelling equilibria and kinetics. In *Responsive Gels: Volume Transitions*, ed. Dušek, I. K., Berlin, Heidelberg, Springer Berlin Heidelberg, **pp.** 233–267.

74. Singhvi, G. and Singh, M. (2011). Review: in-vitro drug release characterization models, *Int. J. Pharm. Stud. Res.*, **2**(1), pp. 77–84.

75. Smela, E. (1999). A microfabricated movable electrochromic "pixel" based on polypyrrole, *Adv. Mater.*, **11**(16), pp. 1343–1345.

76. Smela, E. (2003). Conjugated polymer actuators for biomedical applications, *Adv. Mater.*, **15**(6), pp. 481–494.

77. Smela, E., Inganäs, O. and Lundström, I. (1995). Controlled folding of micrometer-size structures, *Science*, **268**(5218), pp. 1735–1738.

78. Soppimath, K. S., Kulkarni, A. R. and Aminabhavi, T. M. (2001). Chemically modified polyacrylamide-g-guar gum-based crosslinked anionic microgels as pH-sensitive drug delivery systems: preparation and characterization, *J. Control. Release*, **75**(3), pp. 331–345.

79. Srinivasan, G. (2006). Biosensing and drug delivery by polypyrrole, *Anal. Chim. Acta*, **568**(1–2), pp. 119–125.

80. Suh, Y., Kil, D., Chung, K., Abdullayev, E., Lvov, Y. and Mongayt, D. (2011). Natural nanocontainer for the controlled delivery of glycerol as a moisturizing agent, *J. Nanosci. Nanotechnol.*, **11**(1), pp. 661–665.

81. Svirskis, D., Travas-Sejdic, J., Rodgers, A. and Garg, S. (2010). Electrochemically controlled drug delivery based on intrinsically conducting polymers, *J. Control. Release*, **146**(1), pp. 6–15.

82. Tibbitt, M. W., Dahlman, J. E. and Langer, R. (2016). Emerging frontiers in drug delivery, *J. Am. Chem. Soc.*, **138**(3), pp. 704–717.

83. Tonhauser, C., Alkan, A., Schömer, M., Dingels, C., Ritz, S., Mailänder, V., Frey, H. and Wurm, F. R. (2013). Ferrocenyl glycidyl ether: a versatile ferrocene monomer for copolymerization with ethylene oxide to water-soluble, thermoresponsive copolymers, *Macromolecules*, **46**(3), pp. 647–655.

84. Wichterle, O. and Lim, D. (1960). Hydrophilic gels for biological use, *Nature*, **185**(4706), pp. 117–118.

85. Xiao, W., Zeng, X., Lin, H., Han, K., Jia, H.-Z. and Zhang, X.-Z. (2015). Dual stimuli-responsive multi-drug delivery system for the individually controlled release of anti-cancer drugs, *Chem. Commun.*, **51**(8), pp. 1475–1478.

86. Yamato, M., Konno, C., Utsumi, M., Kikuchi, A. and Okano, T. (2002). Thermally responsive polymer-grafted surfaces facilitate patterned cell seeding and co-culture, *Biomaterials*, **23**(2), pp. 561–567.

87. Zhang, J. and Misra, R. D. K. (2007). Magnetic drug-targeting carrier encapsulated with thermosensitive smart polymer: core–shell nanoparticle carrier and drug release response, *Acta Biomater.*, **3**(6), pp. 838–850.

88. Zhang, X.-Z., Wu, D.-Q. and Chu, C.-C. (2004). Synthesis, characterization and controlled drug release of thermosensitive IPN–PNIPAAm hydrogels, *Biomaterials*, **25**(17), pp. 3793–3805.

89. Brannon-Peppas, L. (1993). Controlled release in the food and cosmetics industries. polymeric delivery systems, *Am. Chem. Soc.*, **520**, pp. 42–52.

90. Graham, N. (1992). Poly (ethylene glycol) gels and drug delivery. In *Poly (Ethylene Glycol) Chemistry*, ed. Harris, J. M., Springer, Boston, MA, pp. 263–281.

91. Liu, L., Kost, J., Fishman, M. L. and Hicks, K. B. (2008). A review: controlled release systems for agricultural and food applications. new delivery systems for controlled drug release from naturally occurring materials, *Am. Chem. Soc.*, **992,** pp. 265–281.

92. Mu, L. and Kong, J. (2007). *Smart Hydrogels, Smart Polymers*, CRC Press, Boca Raton, FL, pp. 247–268.

93. Urban, M. W. (2011). *Handbook of Stimuli-Responsive Materials*, Wiley Online Library.

94. Weber, C., Hoogenboom, R. and Schubert, U. S. (2012). Temperature responsive bio-compatible polymers based on poly(ethylene oxide) and poly(2-oxazoline)s, *Prog. Polym. Sci.*, **37**(5), pp. 686–714.

95. Zeisberger, E., Schönbaum, E. and Lomax, P. (1994). *Thermal Balance in Health and Disease: Recent Basic Research and Clinical Progress*, Birkhäuser Basel.

Chapter 4

Conformational Switching in Nanofibers: A New Bioelectronic Interface for Gas Sensors

Sezer Özenler,[a,*] Müge Yücel,[b,*] and Ümit Hakan Yildiz[a]

[a] *Department of Chemistry, İzmir Institute of Technology, Urla, 35430 Izmir, Turkey*

[b] *Department of Bioengineering, İzmir Institute of Technology, Urla, 35430 Izmir, Turkey*

hakanyildiz@iyte.edu.tr

4.1 Introduction

Switchability is a key mechanism living systems use to respond to an external stimulus in a short period of time. The switching capability relies on a reversible conformational change of the biomacromolecules that enhances the ability of living systems to adapt to the extreme environmental conditions. From primitive single

*Equal contribution by these authors.

Switchable Bioelectronics
Edited by Onur Parlak

ISBN 978-981-4800-89-1 (Hardcover), 978-1-003-05600-3 (eBook)
www.jennystanford.com

cells to complex organisms, all living systems have the ability to respond to and adapt reversibly to the changes in their surroundings, for instance, heat shock in bacteria,[1] change of wettability and/or adhesion skills in gecko and mussels,[2,3] and control and regulation of the transport of ions or molecules (cells).[4] However, artificial materials and devices show irreversible behavior under an external effect—they have an unchangeable and fixed form and function because of an immobilized biomolecule.[5]

Mimicking the switchability properties of a living system is an emerging research topic in the field of bioelectronics since the current demand in sensing requires smart biointerfaces. Switchability is often provided by functional molecules[6,7] or monolayers (self-assembly monolayers),[8] polymers and polymer nanoparticles,[9,10] and proteins.[11]

In this chapter, we present the use of 1D polymer nanofibers exhibiting elastic memory that enables rapid switching. This approach emerges as a new alternative for the utilization of smart and switchable biointerfaces. Poly(vinylidene fluoride) (PVDF) nanofibers are selected as a model system for the application of gas sensing. PVDF nanofibers act as a 1D elastic material yielding reversible width change upon gas adsorption. This property has been exploited to fabricate thin-layer switchable sensing platforms for the analysis of exhaled breath. Switchability relies on the elastic memory of polymer chains. This chapter begins with a discussion of the theory of polymer elasticity. The basics of conformational change in a polymer chain and energy-related shape resistance are introduced in the theory section. In the next section, the correlation between the effect of the chemical environment, the structure of polymers, and switchability is discussed. Single-molecule force spectroscopy (SMFS) is also introduced to characterize the switchability and shape resistance of polymer chains. Next, recent studies are briefly reviewed to support the link between switchability and chain elasticity. In the last section, switchability and sensing behavior of nanofibers are elaborately explained by our results.

4.2 Conformational Change and Energy-Related Shape Resistance of a Single Chain

4.2.1 Theory

Polymers consist of covalently bonded repeating units that cause certain rotational constraints, thereby providing shape-resistant memory or elasticity. The chemical composition of the polymer plays a major role in the chain flexibility, the rotational degree of freedom, and the macroscopic elasticity. The long polymer chains exhibit diverse conformations and configurations, leading to the transformation of the chain position from one to another without breaking the chemical bonds by the simple rotation of the units. However, every molecule is located in space with the lowest possible potential energy and for the conformational change to occur, it is necessary for the energy to exceed a certain value, called the "potential rotation barrier" or the "hindrance potential." When a force exceeding the potential rotation barrier is applied, the polymer chain starts changing conformation to a certain extent, and removing the force induces reorganization of the polymer chain to the initial state.[12]

As explained earlier polymers exhibit conformations ranging from tight coils to highly extended structures, such as rigid rods and rigid helical chains. The parameters and, to some extent, the methods utilized for evaluating the degree of chain flexibility are usually different for "flexible" and "stiff" chains. For flexible polymers the most common parameter in use is Flory's[13–15] characteristic ratio C_∞, defined as

$$C_\infty \equiv \lim_{N\to\infty} \frac{< R^2 >_0}{Nl^2} = \frac{\left[< R^2 >_0 / M\right] m_0}{l^2} \tag{4.1}$$

where $<R^2>_0$ is the unperturbed (theta condition) mean-square end-to-end distance, N is the number of main chain bonds of length l, M is the molecular weight of the polymer, and m_0 is the average mass per main chain bond. C_∞ is a quantitative measure of the impact of the hindered rotation about the main chain bonds and rather fixed bond angles on $<R^2>$. For a freely jointed chain (FJC), which has neither rotational hindrance nor bond angle restrictions, $<R^2>$ is equal to Nl^2.[16–19] Thus, the value of C_∞ is equal to 1 for the FJC and larger C_∞ values are indicative of greater departures from the freely jointed character, that is, diminished flexibility.

The relative contributions of fixed bond angles and hindered rotation can be elucidated by modification of the FJC model. Eyring[20] showed that when N is very large $<R^2>$ can be calculated as

$$< R^2 > = \left(Nl^2\right)\left(\frac{1+\cos\theta}{1-\cos\theta}\right), \tag{4.2}$$

where θ is equal to 180° minus the fixed bond angle. Thus, for a polyethylene backbone $\cos\theta = 0.333$ and $<R^2> = 2Nl^2$ with fixed tetrahedral bond angles for a saturated hydrocarbon backbone being expected to double $<R^2>$. In light of the above, an alternative chain flexibility parameter, which is commonly used, is the conformation (steric) factor σ:

$$\sigma = \left(< R^2 >_0^{1/2}\right)\left(< R^2 >_f^{1/2}\right)^{-1} \quad (4.3)$$

This is obtained from the experimental $<R^2>_0$ value, and the $<R^2>$ value calculated for the freely rotating chain, that is, $<R^2>_f\ \sigma$, provides a measure of the relative increase in the end-to-end distance brought about by hindrances to rotation only. Obviously, C_∞ and σ are related for tetrahedral hydrocarbon backbones by

$$C_\infty = 2\sigma^2 \quad (4.4)$$

As Flory has pointed out,[21] the use of C_∞ is to be preferred over the use of σ. This is because small changes in bond angles can lead to large changes in σ and bond angles are rarely known to within more than a few degrees.

The effect of restricted rotation can be taken into account by introducing an additional term into Eq. 4.2. If the hindering potentials are mutually independent for neighboring bonds and symmetrical,[22–24] $<R^2>$ becomes

$$< R^2 >= (Nl^2)\left(\frac{1+\cos\theta}{1-\cos\theta}\right)\left(\frac{1+ < \cos\phi >}{1- < \cos\phi >}\right) \quad (4.5)$$

Here $<\cos\phi>$ is the average rotation angle ($\phi = 0°$ for trans). The assumptions upon which Eq. 4.5 is based are invalid for many macromolecules. Of great significance in this regard has been the development of the rotational isomeric state (RIS) model.[13,14,25,26] These models generally consider only a few discrete conformers of relatively low energies. Interdependence of bond rotations, asymmetric

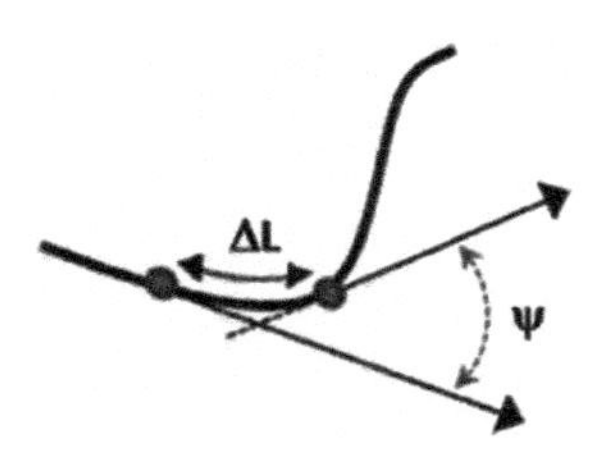

Figure 4.1 A segment of a persistent chain.

rotational potential curves, and finite chain length effects can all be dealt with using modern RIS methods.[13,14]

The Kratky–Porod wormlike chain (WLC) model[27,28] is widely used for describing conformational characteristics of less flexible chains. The polymer is viewed as semiflexible stirring (or worm) of the overall "contour length" L with a continuous curvature. The chain is subdivided into N segments of length ΔL, which are linked at a supplementary angle ψ. The persistence length q (Fig. 4.1) is defined as[29]

$$q = \lim_{\substack{\psi \to 0 \\ \Delta L \to 0}} \frac{\Delta L}{1 - \cos \psi} \tag{4.6}$$

and is thus a measure of the tendency of segments in the polymer chain to "remember" the orientation of adjoining (and other) segments in the chain. WLCs will exhibit conformations ranging from random coils to rigid rods depending on the value of the ratio L/q. Therefore, q provides a measure of chain stiffness. Furthermore, it can be shown[29] that at a large L a WLC becomes Gaussian and q is related to the length of the Kuhn statistical segment l' as

$$q = l'/2. \tag{4.7}$$

Thus, the value of l' is also commonly taken as a measure of chain stiffness. Flory[21] has shown that there is also a simple relationship between the persistence length and the characteristic ratio

$$C_\infty = \frac{2q}{l} - 1 \tag{4.8}$$

Here l is the average bond length and ∞ indicates, as usual, the limiting value for infinite chains. Hence, C_∞ can also be employed in comparing the relative stiffness of WLCs.[15]

Shape-resistant memory, or elasticity, of a polymer chain is a consequence of the interplay between intra- and intermolecular forces acting on each repeating unit.

4.2.2 M-FJC and WLC Models

The modified freely jointed chain (M-FJC) model is utilized to identify the extension of a polymer and produce restoring forces of entropy. In this model a macromolecule is considered as a chain of Kuhn length segments (l_k) that are deformable under stress and statistically independent of each other in Scheme 4.1.[30]

Langevin function describes the extension relevant to the external force:

$$x(F) = \{\coth[(Fl_\mathrm{k})/(k_\mathrm{B}T)] - (k_\mathrm{B}T)/(Fl_\mathrm{k})\}\,(L_\mathrm{c} + nF/K_\mathrm{s}) \tag{4.9}$$

Scheme 4.1 Schematic drawing of the WLC model.

Here, F is the external force, x is the extension of the polymer (end-to-end distance), L_c is the contour length, n is the number of chains stretched, k_B is the Boltzmann constant, T is the temperature, and K_s is the deformability of a segment.

While entropic contribution dominates the elasticity of M-FJC in a low-force region, both its entropy and enthalpy are effective in a high-force region.

WLC is another model that describes the polymer as a homogenous string with fixed bending elasticity. Entropic and enthalpic contributions are combined in the WLC model.[31,32]

The force and extension of the WLC is shown in the following equation:

$$F(x) = [1 - x/L_c)^{-2}/4 - 1/4 + x/L_c]k_B T/l_p \quad (4.10)$$

Here, l_p is the persistence length.

4.3 Force Spectroscopy Applications

SMFS is a technique based on atomic force microscopy (AFM) that allows studies at the molecular level. The force signals cross extension curves and provide new knowledge that cannot be obtained in conventional methods.[30] Here we give a broad overview of SMFS for characterization of switchability properties of single-polymer nanofibers.

SMFS is a technique that quantifies the interaction between a solid substrate hosting polymer chains *physically* or *chemically* adsorbed and the AFM tip. The AFM tip approaches the polymer on the substrate and makes contact with the surface by forming a bridge between polymer and tip. When the AFM tip stretches, the force-versus-extension relation can be deduced to characterize the elasticity of the single chain.[30]

4.3.1 Single-Chain Elongation

The single-chain elongation of poly(ferrocenyldimethy-lsilane) (PFDMS) was studied.[33] Figure 4.2a shows that the force curve of the single-chain elongation curve of PFDMS in tetrahydrofuran (THF) contains one force signal in each curve, which shows the single-chain elongation.

The anchor point of the polymer to the tip or surface is stochastic, and the contour length is varied because the polymer molecular weight is polydisperse. The contour length is scaled by normalizing, which includes dividing the PFDMS force curves by relative extension with the same value of forces. Figure 4.2a indicates the linear proportion between the stretching forces and the relative extension.[33]

According to these force signals all single-polymer chains should superimpose after normalization.[34] The PFDMS polymer chain can be stretched and relaxed continuously by maintaining holding the stretching forces to a point below the rupture point (Fig. 4.2b). This indicates an equilibrium situation, and it is reversible process.[33] As another step in the study an SMFS experiment of oxidized PFDMS and poly(ferrocenylmethylphenylsilane) (PFMPS) in THF buffer has been done. These two polymers have shown different single-molecule-forces curves. In the same way as the preoxidation step, the normalized force-extension (F-E) curves of both polymers superimposed well. And so, there is no hysteresis between trace and

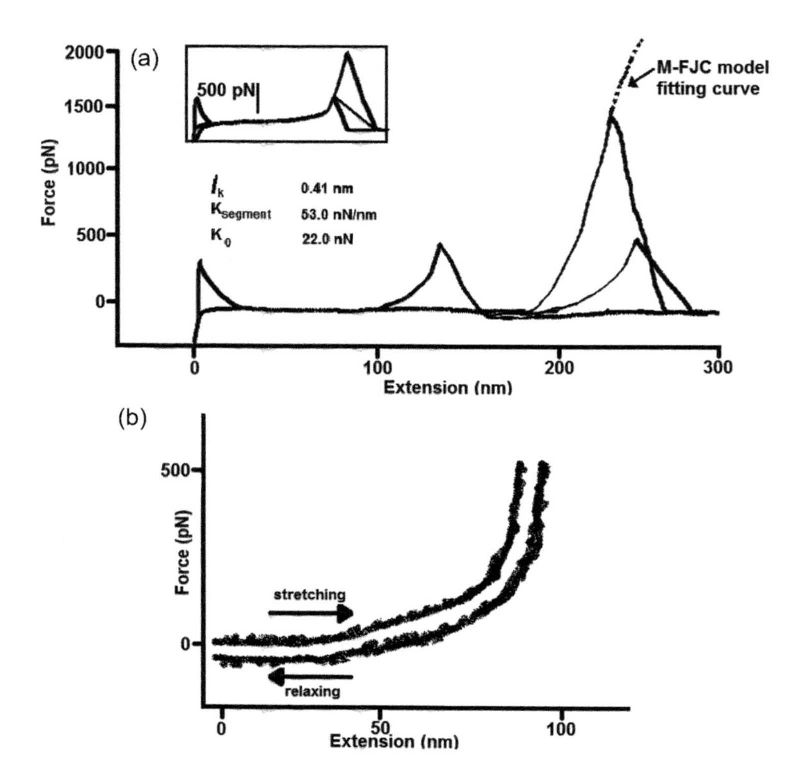

Figure 4.2 (a) Several typical force curves of PFDMS in THF buffer. One of the force curves is fitted by the M-FJC model curve, shown by the dashed line. Inset: Superposition of the normalized force curves. (b) Successive manipulation of a PFDMS single chain, suggesting that the elongation in the experiment is reversible.

retrace curves for either of them after oxidation. The F-E curves of pre- and postoxidized PFDMS in THF buffer overlap in the low-force regime but branch off in the high-force regime, as shown in Fig. 4.3a.

These two forms of a PFDMS single chain show similar entropic elasticity, but the enthalpic elasticity is larger for oxidized poly(ferrocenyldimethylsilane) (o-PFDMS).

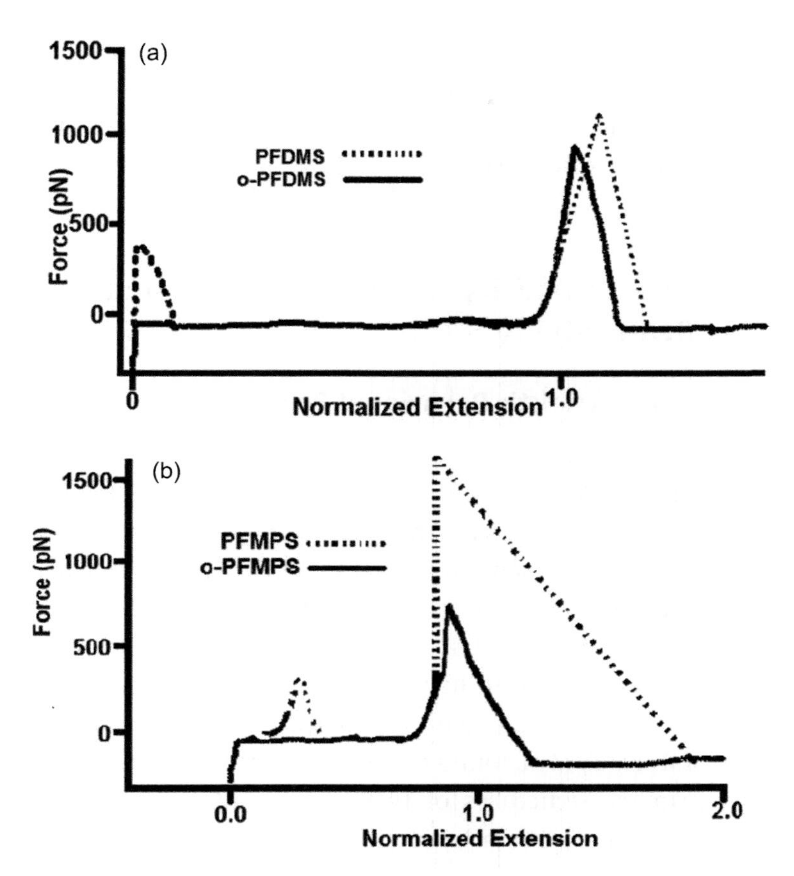

Figure 4.3 (a) Comparsion of the normalized force-extension curves between PFDMS and o-PFDMS in THF buffer. (b) Comparison of the normalized force-extension curves between PFMPS and o-PFMPS in THF buffer.

The PFMPS has a similar tendency in Fig. 4.3b with a steeper form compared to its normal form. The elastic difference between o-PFDMS and oxidized poly(ferrocenyl-methylphenylsilane) (o-PFMPS) can be attributed to the steric effect of the side groups. In oxidized forms Fe^{+3} should interact with the electron-rich phenyl groups stronger than the methyl groups. Enthalpy elasticity of these polymers is similar when they are in their normal form, but the difference can be accurately seen when they are in their oxidized states.

4.4 Chain Elasticity, Shape Resistance, and Switchability

4.4.1 Characteristic Behavior of Polysaccharides

Polysaccharides are fundamental compounds of the living systems and known as the first biological polymers. The glucopyranose ring, with the most stable chair conformation, transforms itself into a chair-boat conformation when it passes over an energy barrier with the influence of an external force. α-(1,4) and β-(1,3) linkages of glucose residues are shown to have an influence over transitional energy during the elongation of polysaccharides.[30] Li et al. and Marszalek et al. found that the little difference between the primary structures of two isomers CM-amylose and CM-cellulose causes a huge difference in their elongation properties.[35,36]

CM-amylose shows a shoulder-like plateau on its force curve at about 300 pN. There is 0.08 nm elongation and 7.3 kT of required energy for conformational transition is observed per glucose residue.[35] CM-cellulose has a sudden increase in its force curve and shows no plateau.

The mechanical property of the polysaccharide chain is intensely affected by different linkages; each successive β-(1,4)-linked glucose residue can readily flip 180° to an extended conformation by an external force effect; however, α-(1,4)-linked glucose residues readily adopt a chair-boat transition in order to achieve an extended conformation during the elongation process. CM-amylose has its authentic property of conformational transition. Marszalek et al. also pointed out that the range between two glycosidic oxygen molecules that determines the length of a monomer of polysaccharide is likely to vary with the pyranose ring. Stretching the polysaccharide chain seems to help conformations and provides further separation of glycosidic oxygen molecules parallel to the extension of the molecule.[35,36] Cellulose has β-(1,4) linkages. In their ab initio calculations glycosidic oxygen vector (O,O4) is already at its maximum value in the chair conformation. Therefore, a cellulose molecule not showing a plateau region in its force curve indicates the conformational transition of the ring structure. These results correspond to the known rigidity of cellulose.

Xu et al. performed a control experiment with a set of carrageenan-bearing oxygen bridges and oxygen bridges not bearing carrageenan. They worked with the λ-carrageenan, κ-carrageenan, and *L*-carrageenan, which are identical in primary structure at the 1,3-linked β-D-pyranose ring of the repeating unit. However, κ-carrageenan and *L*-carrageenan have an oxygen bridge over 1,4-linked α-D-galactopyranose in their other repeating part. λ-Carrageenan showed a shoulder-like plateau as a sign of the conformational transition process, but this process was prohibited by the effect of an additional oxygen barrier in κ-carrageenan and *L*-carrageenan (Fig. 4.4).[37]

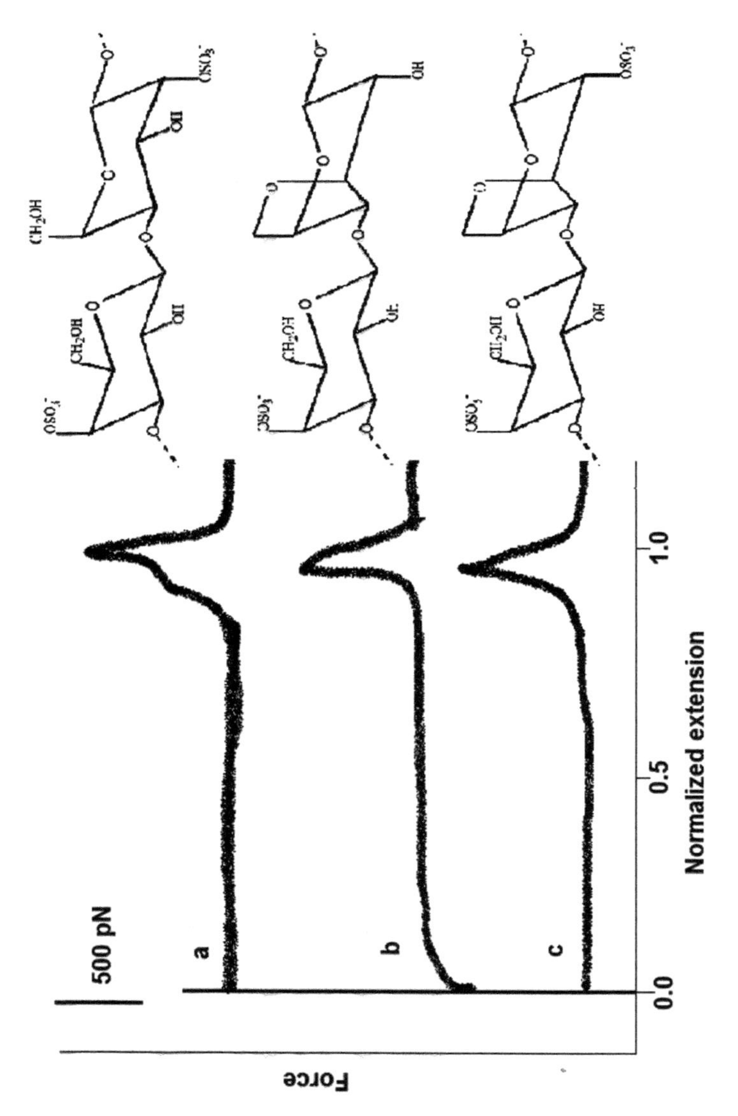

Figure 4.4 Normalization of the force curves obtained on (a) λ-, (b) κ-, and (c) *L*-carrageenan.

4.4.2 Effect of Small-Molecule Polymer Interaction on Switchability

It has become easy to study small molecule–polymer interactions in changing buffer by applying the SMFS technique at the solid-liquid interface in a liquid cell. Alternation of single-chain elasticity in the presence of small molecules proves the interaction between molecules and polymers.[30]

Wang et al. made a comparative study of the single-chain elasticity of PDMA and poly[2-(diethylamino)ethyl methacrylate] (PDEA) with SMFS. These two polymers have the same backbone structure with varying substitutions in *N*-linked groups. Various buffer conditions, that is, aqueous NaOH and aqueous urea solutions of 2 M and 8 M, were applied. The contour lengths of PDMA varied due to the polydisperse nature of the polymer and the uncontrolled stretching point of the AFM cantilever. The F-E curves of the polymer for different contour lengths were normalized, and the superimposed force curves are shown in Fig. 4.5a.[38]

It is observed that the low stretching force and rupture with no hysteresis between stretching and relaxing processes refers to the equilibrium condition. Wang et al. used the M-FJC model to describe the elasticity of a single PDMA polymer chain. Figure 4.5b shows the similar results of the PDEA sample experiments. However, a single PDEA chain is observed to be stiffer in the high-force region. This difference is attributed to the effects of side groups on the polymer backbone. When the concentration of aqueous urea was higher the stiffness of the PDMA increased. This is attributed to the formation of hydrogen bonds in PDMA. To understand the mechanisms of the interaction between

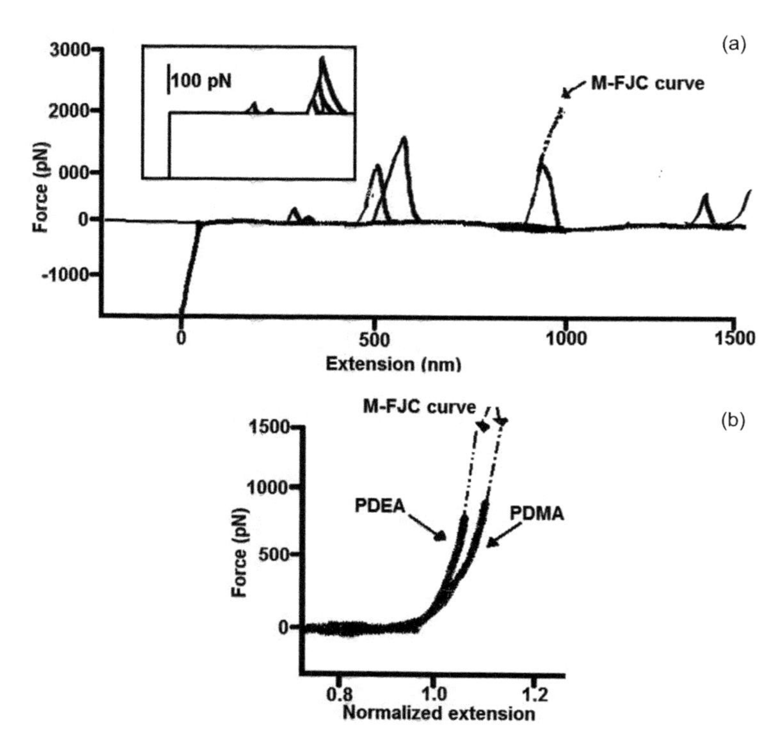

Figure 4.5 (a) Several typical force-extension curves of PDMA in an aqueous solution (pH 9), one of which is fitted by a modified FJC (M-FJC) curve. The fit parameters are l_K (1.3 nm, K segment) 12,000 pN/nm. The normalized force curves are superimposed and plotted in the inset. (b) Comparison of modified FJC fits and normalized force curves of PDMA and PDEA in aqueous solutions.

PDMA and urea, they measured the infrared (IR) spectra of PDMA and 8 M mixture and the 2 M urea mixture. In the case of the 2 M urea solution, Fourier-transform infrared (FTIR) spectra show similar results.

Figure 4.6a shows the similar results for the PDEA tests. The results of the dependence of the urea concentration of the PDMA and PDEA single-chain elasticity in 2 M

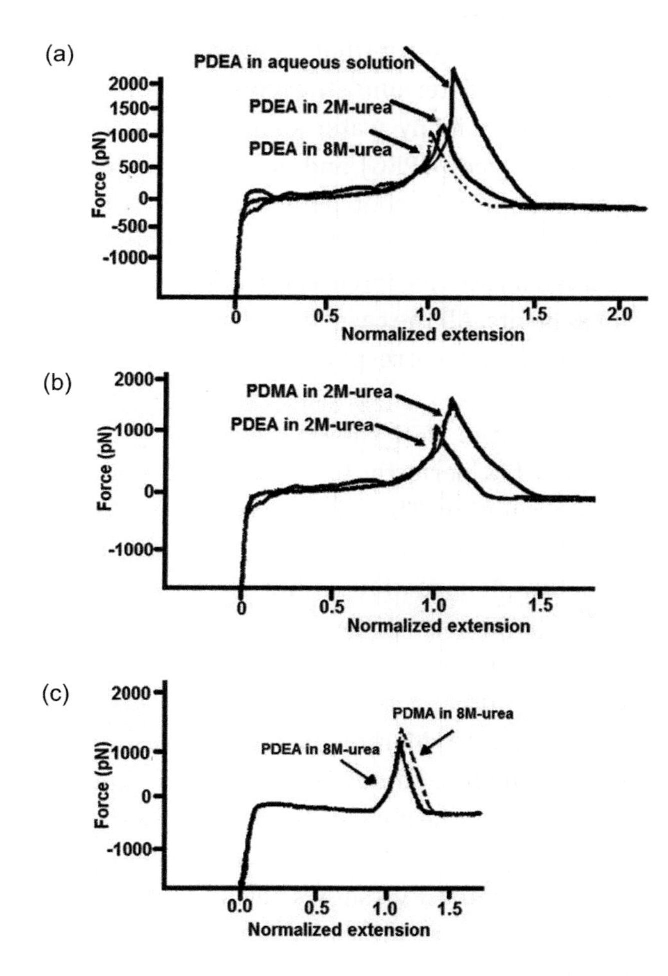

Figure 4.6 Comparison of normalized force curves of (a) aqueous solution and 2 M and 8 M urea buffer solutions for PDEA (b) PDMA and PDEA in 2 M urea solution (c) PDMA and PDEA in 8 M urea solution.

solution are compared in Fig. 4.6b. In the 8 M concentration of urea buffer solution, the chains show no discrepancy (Fig. 4.6c).

These results suggest that sufficiently high concentrations of urea determine the elasticity of the single-polymer chain. In many water-soluble polymers, such as poly(ethylene glycol) (PEG) and poly(vinyl alcohol) (PVA), hydrogen bonds are the controlling factor.[39,40] Oesterhelt et al. performed an experiment to elongate the PEG molecule and observed a resistive force as the elongation function in different solvents. All the cases have given a fully reversible molecular response corresponding to the thermodynamic equilibrium.

The hexadecane PEG shows the ideal entropy spring behavior and fits perfectly to the definition of FJC in the water-based solution. The force curves show a remarkable deviation in the middle force region, which indicates suprastructure deformation within the polymer. A nonplanar suprastructure formed by PEG and stabilized by water bridges in an aqueous solution that is predicted through ab initio calculations agrees well with the analysis based on elastically coupled Markovian two-level systems, and the obtained binding free energy is 3.0 ± 0.3 KT.[39] Also Li et al. investigated PVA at the nanoscale with SMFS and found that the elastic properties of PVA molecules scale linearly with their contour lengths. PVA shows deviation at force spectra, and this deviation may indicate a suprastructure in PVA in NaCl solution.[40]

Liu et al. investigated PVPr single-chain nanomechanical properties in water, ethanol, and THF solvents using the SMFS technique. The F-E curve of PVPr is noticeably deviated in water due to the provided water bridge between the carbonyl groups of pyrrolidone rings. The single chain requires more energy because this water bridge

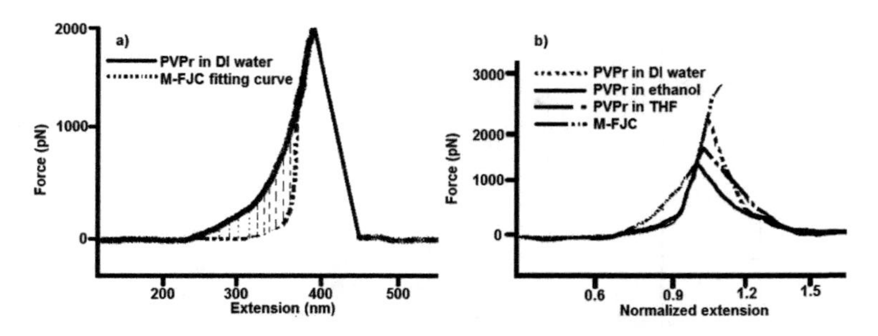

Figure 4.7 (a) Typical force curve of the elongation of a single PVPr chain in DI water. The dotted line is the fitting curve on using the M-FJC model. Fit parameters are l_k 0.63 nm and K_0 41,580 pN. (b) Comparison of the normalized force curves of the elongation of a single PVPr chain in DI water, ethanol, and THF. Fit parameters are l_k 0.31 nm and K_0 38,000 pN.

needs to break for a specific amount of stretch of PVPr (Fig. 4.7).

Liu et al. also studied the PVPr-I_2 force signal in the KI and KI_3 aqueous solution. While the elastic properties of KI_3 were significantly influenced by specific interaction, I_2 and I^- did not influence and this effect is analyzed by using IR. The stretched and relaxed motions repeatedly applied without rupture did not show any hysteresis (Fig. 4.8).[41]

4.4.3 Polymer-Solvent Interaction

In general, the interactions of the polymer with the appropriate solvent result in either dissolution or swelling. In the case of flexible polymer chains, units or segments of the chains can displace with solvent molecules. The dependence of the interaction energy on the intermolecular distance explains why loosely packed polymer molecules have weaker interchain interaction and increased polymer solubility.[12]

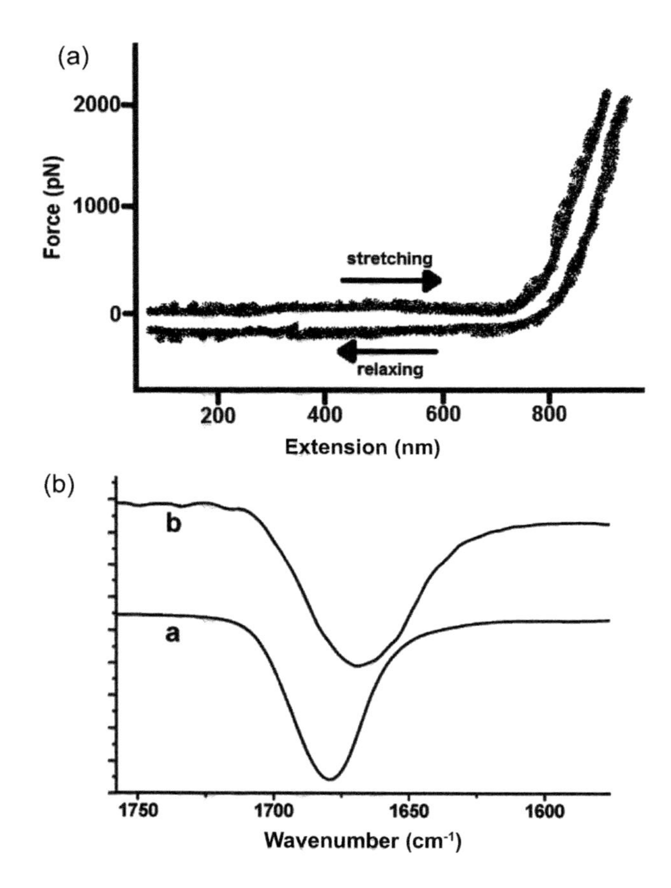

Figure 4.8 (a) Stretching and relaxing traces of the identical PVPr chain in aqueous KI_3 solution. For clarity the two curves are offset. (b) FTIR spectra of PVPr and its mixture with KI_3 at 1750–1600 cm^{-1}: curve a, PVPr film; curve b, PVPr film under the existence of excess KI_3.

In solvents of good thermodynamical characteristics polymer/segment-solvent interactions are the most active and the chain will expand to improve these interactions while reducing the polymer-polymer interaction excluded volume (EV) effect. In contrast in a solvent with poor thermodynamical features, the chain contracts to minimize the impractical polymer-solvent interactions.[15]

Luo et al.[42] were the first ones to discover the notable influence of the EV effect on single-chain mechanics of PEG. They used four different organic solvents, including tetrachloroethane (TCE, $C_2H_2CI_4$), *n*-nonane (C_9H_{20}), *n*-dodecane ($C_{12}H_{26}$), and *n*-hexadecane ($C_{16}H_{34}$), as liquid media in their study. In the experiment with relatively the smallest-size-solvent TCE they observed typical F-E curves, and with normalization superposed F-E curves were obtained, which is a sign of single-chain elasticity of PEG. The plateau that was observed in aqueous solution–based normalization F-E curves is not presented in TCE. The freely rotating chain (FRC) model is the proper way to describe chain elasticity of PEG as it is considered to be a flexible polymer and all C–C and C–O bonds rotate freely. The chain structure of PEG corresponding to the quantum mechanical–freely rotating chain (QM-FRC) model calculation results suggests that the PEG chain is not influenced by the EV interaction of TCE molecules. *N*-hexadecane, $\sim$7 times larger than TCE, is used in experiments also. In Fig. 4.9 a comparison of F-E curves of PEG in TCE and *n*-hexadecane is shown.

Lower energy requirement for the elongation of the PEG chain in an *n*-hexadecane solvent than the TCE obtained is a result of the molecular size difference. Figure 4.10 shows the interaction effect on the polymer chain conformation schematically. Because of the large range in size scale between TCE and *n*-hexadecane two other organic solvents, *n*-nonane (C_9H_{20}) and *n*-dodecane ($C_{12}H_{26}$), were also investigated.

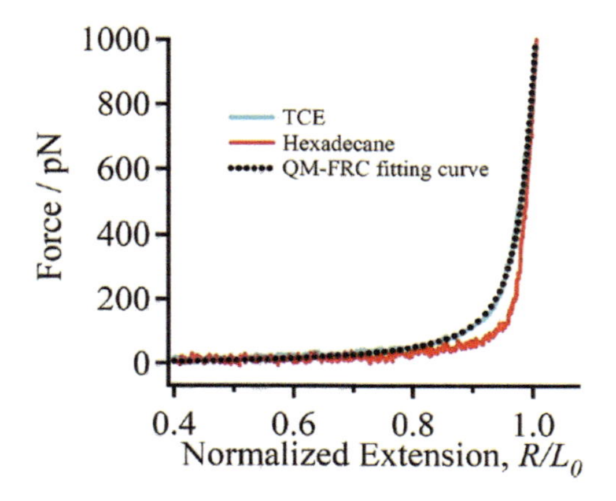

Figure 4.9 Comparison of the single-chain F-E curves of PEG obtained in hexadecane (cyan) and TCE (red). The dotted line is the QM-FRC fitting curve at $l_b = 0.146$ nm (black).

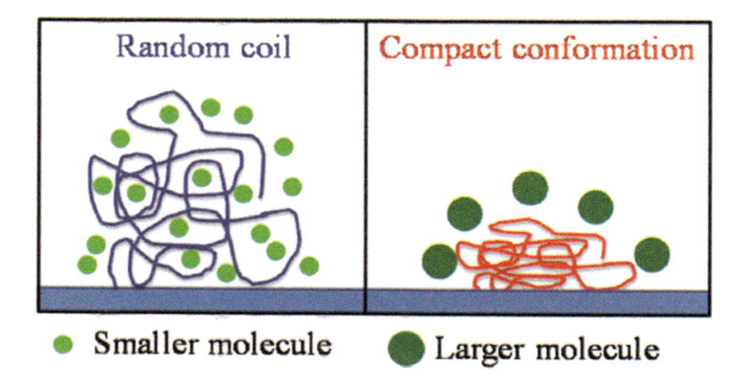

Figure 4.10 The schematic of the EV effect on the conformation of a polymer chain. The small and large spheres represent the solvent molecules of different sizes.

Figure 4.11 shows single-chain F-E curves obtained from four tested solvents. As these results indicate, a nonpolar solvent can be evaluated in two classes: small-sized organic solvents in which PEG elasticity stays novel and middle-sized solvents giving F-E curves different from those of the

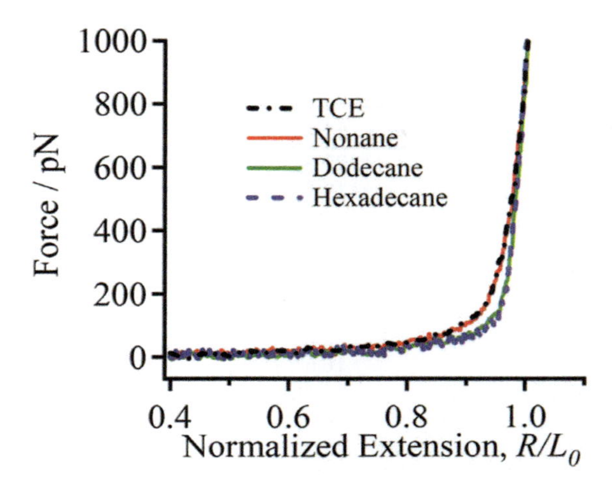

Figure 4.11 The normalized single-chain F-E curves of PEG obtained in organic solvents with different molecular sizes.

small-sized solvents due to their EV effect on the single-chain mechanics of PEG.[42]

Polymers and low-molecular-weight solvents should have close affinity (hydrogen or donor-acceptor bonds) with each other, and the polymer swells before it dissolves, that is, the low-molecular-weight liquid is absorbed and inevitably changes occur in the polymer structure leading to mass and volume increase.[12] A swelling occurs not only in the liquid phase but also in the gas phase, and the swollen polymer in the liquid also reacts with the gas phase of the same liquid. Even though the swelling in the gas phase is slow, it is equal to the same equilibrium state as the swelling degree in the liquid phase.[12] The measurement of the swollen polymer can be monitored gravimetrically or volumetrically.

$$\alpha = \frac{V - V_0}{V_0} \tag{4.11}$$

Table 4.1 Hildebrand parameters of selected solvents and polymers

Solvent	**δ [(MPa)$^{1/2}$]**	**H-bonding tendency**
Acetone	20.3	Medium
Butylenes (iso)	13.7	Poor
Ethyl alcohol	26	Strong
Methanol	29.7	Strong
Toluene	18.2	Poor
Water	47.9	Strong
Polymer		
Poly(ethylene)	16.6	
Poly(vinyl alcohol)	25.78	
Poly(vinyl chloride)	19	
Poly(vinylidene fluoride)	23.2	

Here, α is equal to the swelling rate, V is the volume of the swollen polymer, and V_0 is the volume of the unswollen polymer.

In a study that examined the mechanical behavior of a single-polymer chain in the presence of a nonsolvent in the liquid environment, Horinaka and his colleagues used a polystyrene (PS) chain in a mixture of different volumes of dioxane and methanol. According to their study in good and theta solvents F-E curves show an FRC-like tendency and good reproducibility. Nonsolvent-based experiments gave such results as F-E curves dependent on the extension speed, for example, at a relatively lower speed of 200 nm/s FRC-like was behavior observed while saw-toothed curves were obtained at 2000 nm/s. The shape of the saw-toothed curves was varied due to different measurements. A force relaxation was also seen under a fixed extension distance with 2000 nm/s extension. The mechanical behavior in the nonsolvent points to an inhomogeneous deformation in the PS chain due to a reduction in the chain mobility.

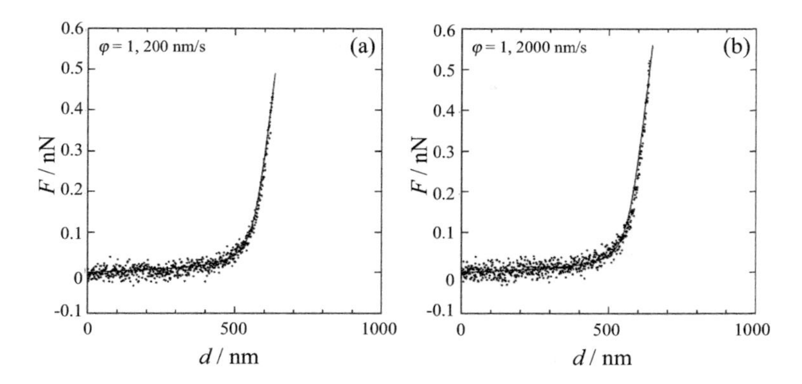

Figure 4.12 Force-extension curves for a PS chain in a liquid of $\varphi =$ 1. The extension rate was (a) 200 nm/s and (b) 2000 nm/s. The solid lines are fitted curves by the m-FJC model with the same parameters.

Because dioxane is a good solvent for PS and methanol is not, $\varphi = 1, 0714, 0.35$, and 0 dioxane volume mixtures were used. $\varphi = 0.714$ is reported to be a theta solvent for PS at 25°C.[43] $\varphi = 0.35$ and $\varphi = 0$ are nonsolvents. Experimental work suggests that at $\varphi = 1$ the curve profile is independent of the extension speed (Fig. 4.12). In $\varphi = 0.35$ there is an obvious difference in F-E curves due to the extension speed, the FJC-like nature at 200 nm/s, and the saw-toothed shape at 2000 nm/s in Fig. 4.13. This data can be expressed in general as follows: the deformation is homogenous under the conditions of small deformation rate and chain motion time characteristic of products, but it is inhomogeneous when the product is large enough.[44]

Latent solvents exhibit little or no solvent properties at room temperature in polymer systems. However, they can become solvents at high temperatures or with appropriate solvent mixtures. For instance, acetone is used as a latent solvent for the PVDF polymer.[45]

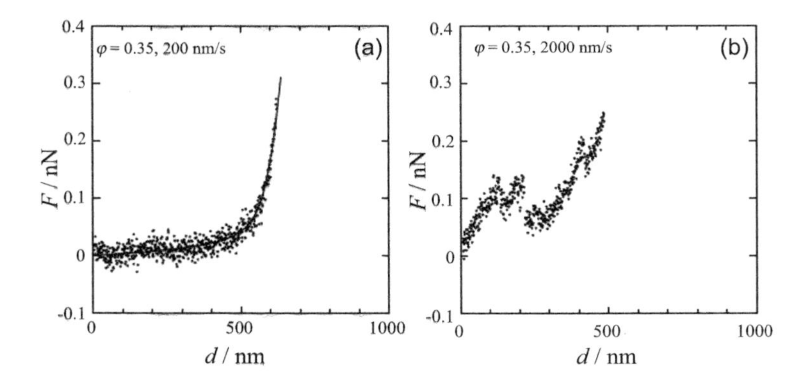

Figure 4.13 Force-extension curves for a PS chain in a liquid of $\varphi =$ 0.35. The extension rate was (a) 200 nm/s and (b) 2000 nm/s. The solid line in (a) is a fitted curve by the m-FJC model.

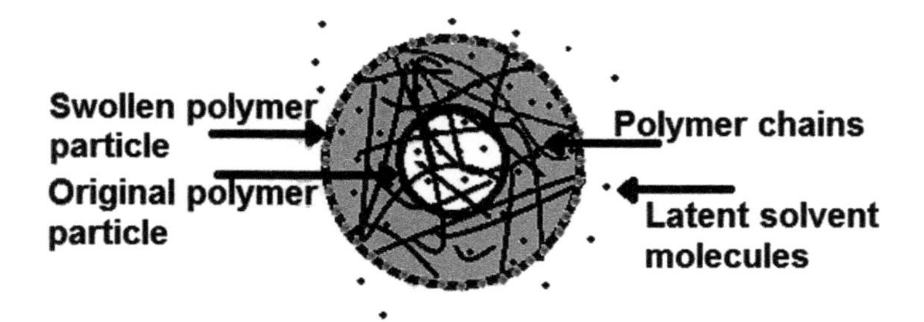

Figure 4.14 Schematic depiction of the swelling of a polyvinyl fluoride particle by its latent solvent.

As a good illustration, in Fig. 4.14, poly(vinyl fluoride) (PVF) films are obtained from the dispersion of PVF using propylene carbonate or other latent solvents (especially ketones), such as N-methyl pyrrolidone (4.1 D), γ-butyrolactone (4.27 D), sulfolane (4.8 D), and dimethyl acetamide (3.7 D). PVF is insoluble at high temperatures with its latent solvent; instead it swells with the diffusion of solvent particles.[46,47]

4.5 Switchable Nanofiber Sensing Platform

Polymer fibers/nanofibers are processed by four techniques: coaxial flow systems, wet-spinning, melt-spinning, and electrospinning.[48] Electrospinning is the commonly used method due to its simplicity and compatibility with all polymers.[49] A simple electrospinning setup consists of a pump, a power source, and a conductive collector. A solution of the desired product is drawn into a syringe, and then it is placed in a pump connected to the device. The pump allows adjustable flow through the nozzle from the syringe. When the solution comes to the nozzle, a drop of the solution accumulates at the tip of the nozzle. Afterward, an increasingly high electrical field is applied. Up to a critical voltage, the drop can be held by its own surface tension.[50] The electrical field electrostatically charges the surface of the drop. When the voltage attains the critical value, the liquid drop begins to elongate and transforms into a conical shape. The liquid polymer solution elongates through the collector, and in the meanwhile, the solvent evaporates.[51] Finally, solid polymeric fibers are collected on the collector. A high yield of nanofibers is obtained by low-cost processing steps. Textiles, substrates, and scaffolds made of nanofibers are becoming common building blocks of cutting-edge-technology devices.

In recent years, several studies have been performed to characterize PVDF fiber formation. Fiber formation relies on parameters such as solution concentration and applied potential. Figure 4.15 illustrates the typical effect of solution concentration and applied potential; as the concentration increases to 25%, fibers of defined shapes are obtained.[52] A concentration decrease causes bead

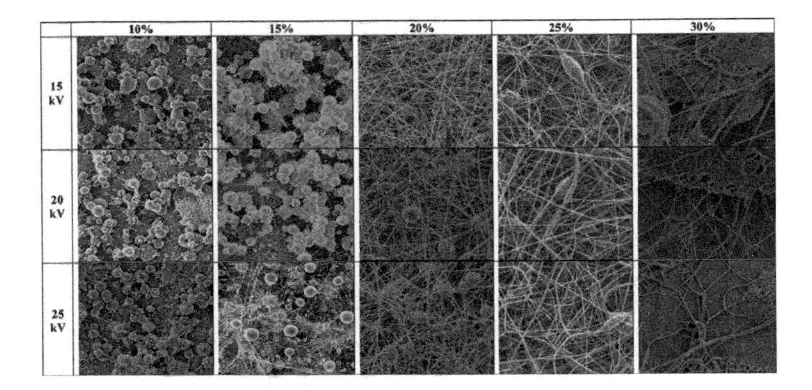

Figure 4.15 Morphology of PVDF nanofibers (polymer flow rate, 1.5 mL/min; tip-to-collector distance, 15 cm) showing the presence of beads and inhomogeneities. Nanofibers displaying minimum defects were obtained at 20 kV, 25 wt%.

formation. The applied potential causes a marginal change in the fiber diameter. Pellerin and Richard-Lacroix have discussed the emergence of characterization techniques that enable study at the single-fiber level and, therefore, study of the structures and properties of electrospun nanofibers with their molecular orientations.[53] PVDF fibers are readily fabricated by electrospinning from their viscous solution. In our studies, a bicomponent solvent system (acetone-DMF) was used for fiber processing.[54] Scanning electron microscopy (SEM) characterization of PVDF nanofibers is shown in Fig. 4.16. The fiber diameters range from 60 nm to 150 nm, and the average fiber diameter (AFD) was found to be 65 nm. The fiber density of the final electrospun product was calculated to be approximately 87.44%. The diameter and density of fibers are vital properties for the fabrication of polymer sensor platforms. The elasticity of the fiber mat is a critical parameter to control the switchability

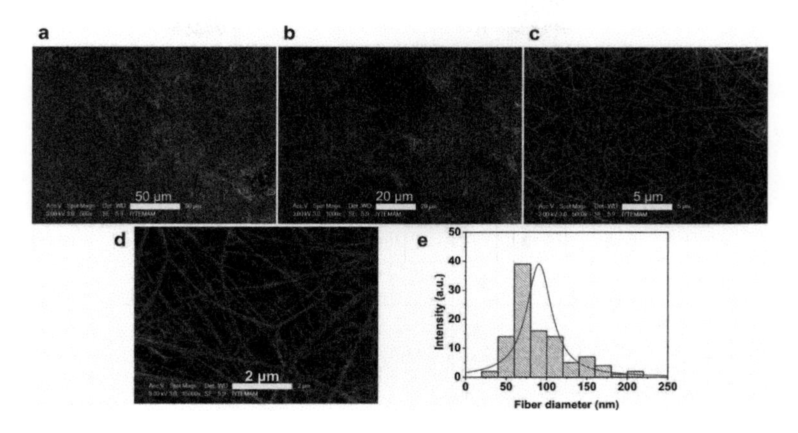

Figure 4.16 SEM images of PVDF fibers magnified (a) 500 times, (b) 1000 times, (c) 5000 times, and (d) 15,000 times and (e) AFD distribution of fibers from (d).

Table 4.2 Properties of (1) pristine PVDF fiber and (2) PVDF fiber exposed to acetone and (3) PVDF fiber exposed to toluene

	Normalized extension work	Normalized Young's modulus	Relative stiffness	Relative deformation
1	1.00	1.00	1.00	1.00
2	0.33	0.44	0.79	1.02
3	0.66	0.69	0.86	1.09

properties. As the elastic modulus of the fiber increases, the volume and shape of the fiber changes drastically, which causes displacements and entanglements. Reversible transformations in the conformation and position of fibers by a stimulus govern the switchability as well as response of fibers.

The elasticity, deformation, adhesion and peak forces, stiffness, and Young's modulus of the polymer fiber were characterized by the peak force mode of AFM. Figure 4.17

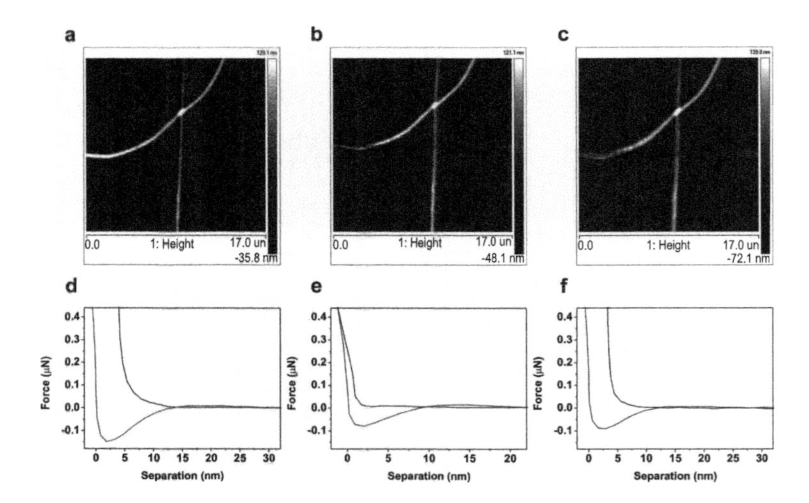

Figure 4.17 Micrographs of (a) PVDF fiber, (b) acetone-exposed PVDF fiber, and (c) toluene-exposed PVDF fiber and force-separation curves of (d) PVDF fiber, (e) acetone-exposed PVDF fiber, and (f) toluene-exposed PVDF fiber.

illustrates force separation curves and AFM micrographs under ambient conditions as well as acetone and toluene atmospheres.

AFM micrographs exhibits marginal shape variations contrary to peak force analysis, indicating substantial deviations in Young's modulus as well as stiffness. Exposure to acetone lowers Young's modulus of PVDF by 56%, while exposure to toluene causes ~30% decrease in Young's modulus. The decrease in Young's modulus of the single fiber indicates that there is a much steeper deformation in PVDF when acetone is applied, due to polymer-solvent molecule interactions. The PVDF-acetone interaction causes a softening on the fiber surface, lowering stiffness. As acetone is removed, the polymer solidifies rapidly to the initial conditions. The reversible softening-

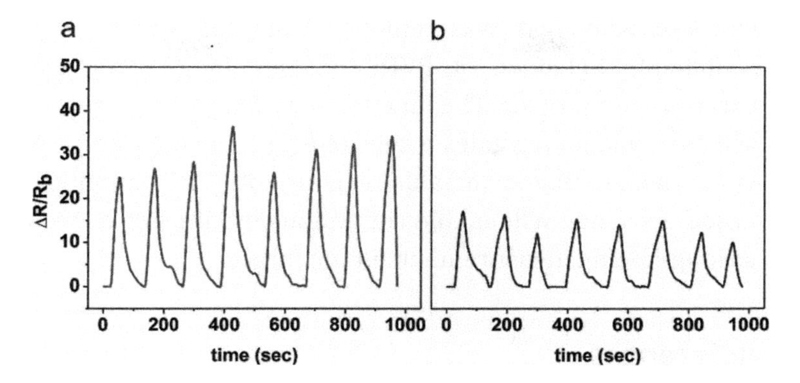

Figure 4.18 Typical resistivity response ($\Delta R/R_b$) of a PVDF sensor to (a) acetone vapor and (b) toluene vapor.

solidifying processes provide switchable shape transformation to PVDF. The same behavior was not observed for PVDF-toluene due to the low extent of interaction of polymer and solvent.

Switchable polymer-solvent interactions hold great promise for gas sensors. PVDF-carbon nanotube (CNT) composites were validated for acetone and toluene sensing. Figure 4.17 illustrates the typical resistivity response of a PVDF-based sensor platform at the conductive interface. Here the platform was exposed to acetone vapor and toluene vapor separately. When the gas molecules are adsorbed on the sensor, resistivity increases and electron flow in the CNT network is obstructed by molecules. As discussed above, PVDF was subjected to substantial swelling and conformational changes due to acetone exposure. The conductive layer (CNT) was displaced significantly by PVDF-acetone interactions. As illustrated in Fig. 4.18a, acetone causes a resistivity change of ~20% while toluene causes a change of ~10%. The initial conductivity was restored when gas was removed from the system. The switchable behavior of PVDF enables long-lasting gas exposure monitoring. The functionalization of fibers by gas-selective agents provides a selective gas sensing interface. We anticipate that switchable response of PVDF nanofiber-based electrode will enable continuous monitoring as well as long-lasting measurements in the future.

References

1. Jian, H., et al. (2016). Global transcriptome analysis of the heat shock response of the deep-sea bacterium Shewanella piezotolerans WP3, *Mar. Genomics*, **30**, pp. 81–85.
2. Gao, H., et al. (2005). Mechanics of hierarchical adhesion structures of geckos, *Mech. Mater.*, **37**(2–3), pp. 275–285.
3. Lee, H., Lee, B. P. and Messersmith, P. B. (2007). A reversible wet/dry adhesive inspired by mussels and geckos, *Nature*, **448**(7151), pp. 338–341.
4. Sepúlveda, F. V., et al. (2015). Molecular aspects of structure, gating, and physiology of pH-sensitive background K_{2P} and kir K^+-transport channels, *Physiol. Rev.*, **95**(1), pp. 179–217.
5. Parlak, O. and Turner, A. P. F. (2016). Switchable bioelectronics, *Biosens. Bioelectron.*, 2016. **76**, pp. 251–265.
6. Qi, Q., et al. (2015). Reversible multistimuli-response fluorescent switch based on tetraphenylethene-spiropyran molecules, *Chemistry*, **21**(3), pp. 1149–1155.
7. Dabrowa, K., Niedbala, P. and Jurczak, J. (2016). Engineering light-mediated bistable azobenzene switches bearing urea d-aminoglucose units for chiral discrimination of carboxylates, *J. Org. Chem.*, **81**(9), pp. 3576–3584.
8. Kashi, M. B., et al. (2016). Silicon–SAM–AuNP electrodes: Electrochemical "switching" and stability, *Electrochem. Commun.*, **70**, pp. 28–32.

9. Becherer, T., et al. (2015). In-depth analysis of switchable glycerol based polymeric coatings for cell sheet engineering, *Acta Biomater.*, **25**, pp. 43–55.

10. Yuan, C., et al. (2015). Supramolecular assembly of crosslinkable monomers for degradable and fluorescent polymer nanoparticles, *J. Mater. Chem. B*, **3**(14), pp. 2858–2866.

11. Nagel, G., et al. (2005). Channelrhodopsins: directly light-gated cation channels, *Biochem. Soc. Trans.*, **33**(4), pp. 863–866.

12. Tager, A. (1978). *Physical Chemistry of Polymers*, MIR, Moscow.

13. Flory, P. J. (1969). *Statiscal Mechanics of Chain Molecules*, Interscience, New York.

14. Flory, P. J. (1974). Foundations of rotational isomeric state theory and general methods for generating configurational averages, *Macromolecules*, **7**(3), pp. 381–392.

15. Xu, Z., et al. (1995). Structure/chain-flexibility relationships of polymers. In *Physical Properties of Polymers*, Springer Berlin Heidelberg, pp. 1–50.

16. Kuhn, W. (1934). Über die gestalt fadenförmiger moleküle in lösungen, *Kolloid-Z.*, **68**(1), pp. 2–15.

17. Kuhn, W. (1936). Beziehungen zwischen Molekülgröße, statistischer Molekülgestalt und elastischen Eigenschaften hochpolymerer Stoffe, *Kollloid-Z.*, **76**, p. 258.

18. Kuhn, W. (1939). Molekülkonstellation und Kristallitorientierung als Ursachen kautschukähnlicher Elastizität, *Kolloid-Z.*, **87**, p. 3.

19. Guth, E. and Mark, H. (1934). Zur innermolekularen, statistik, insbesondere bei Kettenmolekülen, *Monatsh. Chem.*, **65**, p. 93.

20. Eyring, H. (1932). The reultant electricmoment of complex molecules, *Phys. Rev. Lett.*, **39**, p. 746.

21. Flory, P. J. (1969). *Statistical Mechanics of Chain Molecules*, Interscience, New York.

22. Benoit, H. (1947). Sur la statique des chains avec interactions et empechements ateriques, *J. Chem. Phys.*, **44**, p. 18.

23. Kuhn, H. (1947). Restricted bond rotation and shape of unbranched saturated hydrocarbon chain molecules, *J. Chem. Phys.*, **15**, p. 843.

24. Taylor, W. J. (1948). Average length and radius of normal paraffin hydrocarbon molecules, *J. Chem. Phys.*, **16**, p. 257.

25. Volkenstein, M. V. (1963). *Configurational Statistics of Polymer Chains*, Interscience, New York.

26. Birshtein, T. M. and Ptitsyn, O. B. (1966). *Conformation of Macromolecules*, Interscience, New York.

27. Kratky, O. and Porod, G. (1949). Röntgenuntersuchung gelöster fadenmoleküle, *Recl. Trav. Chim.*, **68**, p. 1106–1122.

28. Porog, G. (1949). Zusammenhang zwischen mittlerem endpunktsabstand und kettenlange bei fadenmolkülen, *Monatsh. Chem.*, (2), pp. 251–255.

29. Fujita, H. (1990). *Polymer Solutions*, Elsevier, Amsterdam.

30. Mark, J. E. (2007). *Physical Properties of Polymers Handbook*, Vol. 1076, Springer, New York.

31. Smith, S., Finzi, L. and Bustamante, C. (1992). Direct mechanical measurements of the elasticity of single DNA molecules by using magnetic beads, *Science*, **258**(5085), pp. 1122–1126.

32. Senden, T. J., di Meglio, J. M. and Auroy, P. (1998). Anomalous adhesion in adsorbed polymer layers, *Eur. Phys. J. B*, **3**(2), pp. 211–216.

33. Shi, W., et al. (2004). Single-chain elasticity of poly(ferrocenyldimethylsilane) and poly(ferrocenylmethylphenylsilane), *Macromolecules*, **37**(5), pp. 1839–1842.

34. Huge, T. and Seitz, M. (2001). The Study of molecular interactions by AFM force spectroscopy, *Macromol. Rapid Commun.*, **22**, pp. 989–1016.

35. Marszalek, P. E., et al. (1998). Polysaccharide elasticity governed by chair-boat transitions of the glucopyranose ring, *Nature*, **396**(6712), pp. 661–664.

36. Li, H., et al. (1999). Single-molecule force spectroscopy on polysaccharides by AFM–nanomechanical fingerprint of α(1,4)-linked polysaccharides, *Chem. Phys. Lett.*, **305**, pp. 197–201.

37. Xu, Q., Zhang, W. and Zhang, X. (2002). Oxygen bridge inhibits conformational transition of 1,4-linked α-d-galactose detected by single-molecule atomic force microscopy, *Macromolecules*, **35**(3), pp. 871–876.

38. Wang, C., et al. (2002). Force spectroscopy study on poly(acrylamide) derivatives: effects of substitutes and buffers on single-chain elasticity, *Nano Lett.*, **2**(10), pp. 1169–1172.

39. Oesterhelt, F., Rief, M. and Gaub, H. E. (1999). Single molecule force spectroscopy by AFM indicates helical structure of poly(ethylene-glycol) in water, *New J. Phys.*, **1**(1), p. 6.

40. Li, H., et al. (1998). Single molecule force spectroscopy on poly(vinyl alcohol) by atomic force microscopy, *Macromol. Rapid Commun.*, **19**(12), pp. 609–611.

41. Liu, C., et al. (2005). Single-chain mechanical property of poly(N-vinyl-2-pyrrolidone) and interaction with small molecules, *J. Phys. Chem. B*, **109**(31), pp. 14807–14812.

42. Luo, Z., et al. (2016). Effect of the size of solvent molecules on the single-chain mechanics of poly(ethylene glycol): implications on a novel design of a molecular motor, *Nanoscale*, **8**(41), pp. 17820–17827.

43. Brandrup, J., Edmund, E. H. and E. A. Grulke (2003). *Polymer Handbook*, 4th ed., Wiley.

44. Horinaka, J.-i., Takaki, T. and Takigawa, T. (2011). Mechanical behavior of a single polymer chain in a non-solvent, *Polymer*, **52**(24), pp. 5644–5647.

45. Huang, X. (2013). Cellular porous polyvinylidene fluoride composite membranes for lithium-ion batteries, *J. Solid State Electrochem.*, **17**(3), pp. 591–597.

46. David, R. L. (2004). *Handbook of Chemistry*, 85th ed., CRC Press.

47. Ebnesajjad, S. (2013). Preparation and properties of vinyl fluoride. In *Polyvinyl Fluoride*, William Andrew, pp. 25–46.

48. Loes, B. A. S. (2015). *Microfluidcs for Medical Applications*, Vol. 36, Royal Society of Chemistry, UK.

49. Rutledge, G. C. and Fridrikh, S. V. (2007). Formation of fibers by electrospinning, *Adv. Drug Deliv. Rev.*, **59**(14), pp. 1384–1391.

50. Zhao, Z., et al. (2005). Preparation and properties of electrospun poly(vinylidene fluoride) membranes, *J. Appl. Polym. Sci.*, **97**(2), pp. 466–474.

51. Cozza, E. S., et al. (2013). On the electrospinning of PVDF: influence of the experimental conditions on the nanofiber properties, *Polym. Int.*, **62**(1), pp. 41–48.

52. Magniez, K., De Lavigne, C. and Fox, B. L. (2010). The effects of molecular weight and polymorphism on the fracture and thermo-mechanical properties of a carbon-fibre composite modified by electrospun poly (vinylidene fluoride) membranes, *Polymer*, **51**(12), pp. 2585–2596.

53. Richard-Lacroix, M. and Pellerin, C. (2013). Molecular orientation in electrospun fibers: from mats to single fibers, *Macromolecules*, **46**(24), pp. 9473–9493.

54. Yücel, M. (2015). Fabrication of thin layer polymer-based biointerphase for biosensing application, Department of Chemistry, İzmir Institute of Technology.

Chapter 5

Molecularly Imprinted Polymers as Recognition and Signaling Elements in Sensors

Hasan Basan and Hüma Yılmaz
Department of Analytical Chemistry, Faculty of Pharmacy, Gazi University, 06330 Etiler, Ankara, Turkey
basan@gazi.edu.tr

5.1 Introduction

Molecular structures in living organisms have the ability to differentiate foreign molecules through various types of interactions, and this ability is called "molecular recognition." There are a number of natural recognition elements, such as antibodies, enzymes, aptamers, and nucleic acids, used for the detection of target molecules in chemical and biological analyses. Due to the poor chemical, stability, and long-term stability and high cost of these elements, researchers have

Switchable Bioelectronics
Edited by Onur Parlak

ISBN 978-981-4800-89-1 (Hardcover), 978-1-003-05600-3 (eBook)
www.jennystanford.com

focused on investigating alternative synthetic recognition elements that can overcome these limitations. One of the promising approaches is the molecular imprinting technique, and the materials produced are called "molecularly imprinted polymers" (MIPs). MIPs are referred to as "antibody mimics" or "plastic antibodies" because these materials mimic the interactions of their natural counterparts, resulting in high affinity and selectivity.[1]

Differences between antibodies and MIPs are presented in Table 5.1, and as can be seen from the table MIPs have properties comparable with those of their natural counterparts. It should be pointed out that molecular imprinting is a combination of molecular recognition and polymer chemistry or material chemistry principles. Thus, MIPs can be employed in a variety of applications, such as separation; catalysis; and sensors for the detection of drugs in biological samples, toxins, pesticides, food components, etc.[2]

Molecular imprinting is a technique that creates artificial recognition sites in synthetic polymers using molecules or ions as a template. The resulting recognition or imprinted sites, which are complementary to template molecules in size, shape, and the orientation of the functional groups, are generated in the polymeric matrix.[3] MIPs are prepared by the copolymerization of functional and cross-linking monomers in the presence of a template molecule, an initiator, and a porogen. Thereafter, the template is removed by applying a solvent extraction process, generating highly cross-linked 3D cavities having a high affinity toward the target compound. The principle of molecular imprinting is schematically shown in Fig. 5.1.[4] There are two main types of molecular imprinting approaches. The first one is

Table 5.1 Comparison of antibodies and molecularly imprinted polymers

	Antibodies	**Molecularly imprinted polymers**
Shelf life	They can survive several days at room temperature.	They can survive several months to years.
Cost	They are expensive.	They are cheap.
Stability	They show low stability (the antibody can be denatured due to chemical and physical factors).	MIPs show very high resistance to harsh chemical media (high pH values, temperature, and pressure) and are reusable many times without loss of the activity.
Solvent	Use of organic solvents is problematic due to antibody denaturation, but antibodies show high performance in an aqueous medium.	They show excellent performance in organic solvents, but water compatibility is poor.
Binding site homogeneity	Monoclonal antibodies have homogeneous binding sites, but polyclonal antibodies have heterogeneous ones.	MIPs prepared using a noncovalent approach have heterogeneous binding sites; however, MIPs prepared by a covalent approach have homogeneous binding sites.
Affinity	They exhibit high affinity to the target molecule.	The synthesis environment and conditions determine the affinity.

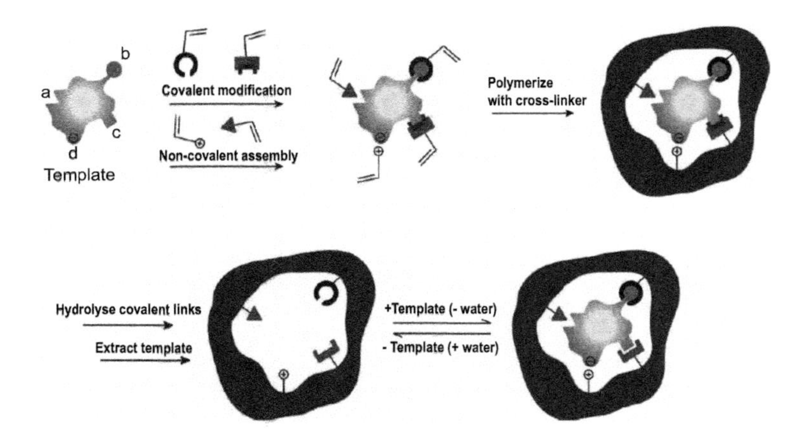

Figure 5.1 Principle of molecular imprinting (reproduced with permission from Ref. [4]).

covalent molecular imprinting, which was introduced by Wulff and Sarhan in 1972.[5] In this approach, functional monomers interact with the functional groups on the template via reversible covalent bonding, resulting in a polymerizable template. After polymerization, these bonds in the covalent character are cleaved via hydrolysis and the template is removed, leaving behind selective binding sites.

During the rebinding assay, these covalent bonds are reformed between the target compound and the functional monomer in the polymeric matrix to selectively remove the target molecule.[6] In covalent imprinting, the functional monomer-template complex is very stable and the functional monomer and the template react in a stoichiometric ratio. Therefore, neither the functional monomer nor the template is used in excess amounts. In addition, the distribution of imprinted sites in the polymeric matrix is homogeneous and higher-affinity binding cavities are generated as a result. On the other hand, since the synthesis of a covalently linked polymerizable monomer-template conjugate is required before imprinting, both conjugate synthesis and removal of the template require labor-intensive chemical processes. Furthermore, due to rebinding in a covalent character, the binding kinetics is low and time consuming. Another molecular imprinting approach was introduced by Mosbach and Arshady,[7] based on noncovalent interactions, such as hydrogen bonding and hydrophobic, π–π, and dipole–dipole interactions. Soon, it became the most widely used molecular imprinting approach due to its simplicity and the availability of a variety of functional monomers and template molecules appropriate for imprinting, compared to the covalent approach. In the noncovalent approach, there is no need for the synthesis of a polymerizable template-functional monomer conjugate. Imprinting is simply achieved by dissolving the functional monomer, cross-linker, and template in a porogen and adding an appropriate initiator into the mixture. After the imprinting process, the template is removed via a solvent extraction process, for example, Soxhlet extraction, producing empty cavities ready for binding. Thereafter, the target compound rebinds with the imprinted sites through noncovalent interactions and high-binding kinetics is obtained. Another advantage of the noncovalent imprinting over the covalent approach is that the number of template molecules and functional monomers available is higher. On the other hand, the main disadvantage of this approach is the nonspecific binding sites generated during the imprinting process because the excess amount of functional monomer should be used to obtain enough number of functional monomer-template complexes to form recognition sites. Therefore, the choice of functional monomer, template, cross-linker, porogen, and initiator is of vital importance in order to create highly selective binding sites.[8] Another parameter is the choice of the polymerization type. Due to its simplicity, radical polymerization is the most widely used polymerization type. Radical polymerization reactions can be done either thermally or photochemically using radicals forming initiators. Since the complex formed between template and functional monomer is more stable at low temperatures, photochemical polymerization usually becomes the method of choice in generating free radicals triggering the polymerization reactions.

Various polymerization methods have been used in the preparation of MIPs, such as bulk, suspension, multistep swelling, and precipitation polymerization. These methods are summarized in Table 5.2, together with their advantages and disadvantages.[9]

Table 5.2 Comparison of different polymerization methods used in the preparation of MIPs

Polymerization type	Advantage	Disadvantage
Bulk	Smooth and easy to prepare	A tedious process in terms of grinding, sieving, and removal of the template; poor performance
Suspension	Small, reproducible particles synthesized	A complicated method; specific surfactant polymers required
Multistep swelling	Monodisperse beads of controlled diameter synthesized	Tough processes and experimental conditions; aqueous emulsions required
Precipitation	Large amounts of homogeneous microsphere particles synthesized	A large amount of template required; a high dilution factor

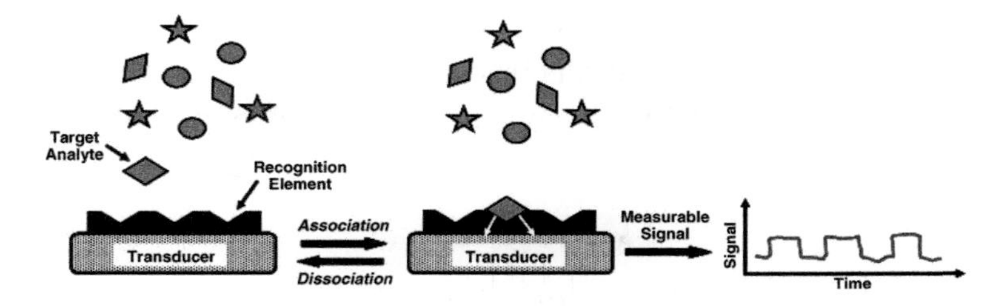

Figure 5.2 Schematic representation of an MIP-based sensor (reproduced with permission from Ref. [10]).

5.2 Applications of the Molecular Imprinting Technique as a Sensor

A chemical sensor consists of a molecular recognition element and a transducer as represented in Fig. 5.2.[3] Molecularly imprinted materials may function as a biomimetic recognition element, and several types of transducers (optical, electrochemical, etc.) can be used to detect the binding event. In designing and developing effective sensors, MIPs are used as a sensor interface due to their ability to selectively recognize analyte molecules and their stability toward chemical and physical effects.

Recently, the combination of recognition or sensing elements and transducers that converts the interaction of a target analyte with a recognition element into an electrical signal has garnered interest in the design and development of sensors.

Electrochemical, optical, and mass-sensitive transducers are the most popular types of transducers used in sensor design. Combinations of these transducers and MIPs are widely used as chemical or biochemical sensors for the detection of target analytes in biological, environmental, food, and drug containing samples. Thus, this chapter focuses on MIP-based electrochemical, optical, and mass-sensitive sensors, with special emphasis on the optical ones.

5.2.1 MIP-Based Electrochemical Sensors

Electrochemical sensors are devices that detect in real time signal change resulting from the interaction of analyte molecules and the recognition element immobilized on the surface of an electrochemical transducer.[11] Since simple procedures and instrumentation are required, they are the largest and oldest group of sensors. They might be amperometric or voltammetric (change of current for an electrochemical reaction with an applied voltage or with time at a fixed potential), potentiometric (change of membrane potential), or conductometric (change of conductance) type, depending on the transduction principle.[12] The first potentiometric MIP sensors using electropolymerized recognition elements were presented by Boyle and coworkers and Vinokurov,[13,14] and sensors for pyrrole, aromatic amines, and substituted phenols were prepared.[15] A similar amperometric MIP sensor was described by Piletsky et al. for the detection of aniline and phenol.[16] The first reported study on integrated conductometric sensors used a phenylalanine analide–imprinted polymer membrane.[17]

In potentiometric sensors, an electrical potential develops at an electrode surface—an ion-selective electrode (ISE) membrane—in contact with the sample solution. The analyte in the sample solution should be in ionic form to generate an electrical potential. In designing an MIP-based

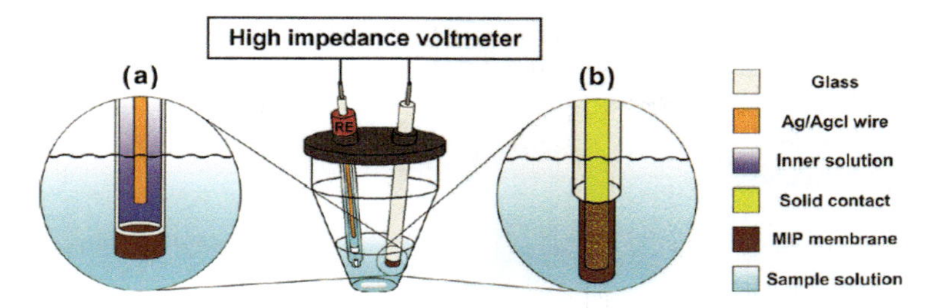

Figure 5.3 Schematic representation of an MIP-based potentiometric sensor system: (a) with an inner reference electrode ISE and (b) with a coated wire ISE. RE: reference electrode (reproduced with permission from Ref. [11]).

potentiometric sensor, the MIP as a recognition element has to be brought into close contact with the transducer surface. Thereafter, a chemical signal is generated as a result of the analyte binding to the MIP and the transducer turns this signal into an electrical output. Then, the potential is measured against a reference electrode while keeping the current at zero and the magnitude of the signal is proportional to the analyte activity in the sample. Here, the MIP-based recognition element serves to enhance the selectivity of the ISE. There are also some reviews in the literature on the recent development of MIP-based potentiometric sensors.[18] An MIP-based potentiometric sensor is schematically represented in Fig. 5.3.

Since conductance measurement is not a selective method, a recognition element, such as an MIP, is used to make the conductometric method selective.[11] Conductometric sensors measure the change in the conductivity of a sample solution due to a chemical reaction, and the percentage of conductance is recorded. A schematic representation of a conductometric MIP–based sensor

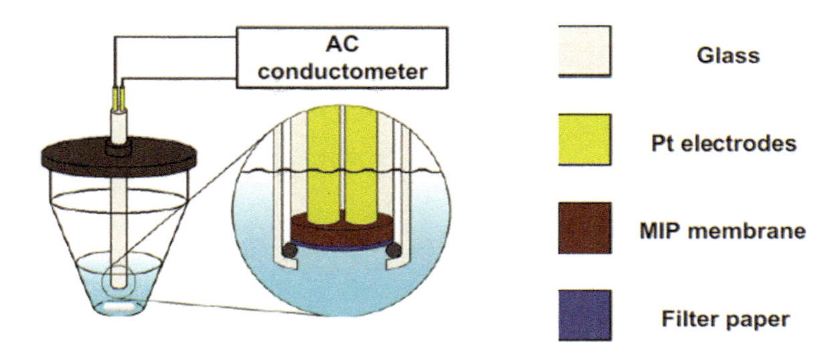

Figure 5.4 Schematic representation of a conductometric sensor device (reproduced with permission from Ref. [11]).

device is given in Fig. 5.4. There are some interesting studies in the literature using MIPs as the recognition element in conductometric sensors.[19–22]

In voltammetric and amperometric techniques, a potential is applied to a working (or indicator) electrode and a reference electrode and then the current is measured. The current is obtained as a result of an electrochemical reduction or oxidation taking place at the working electrode.[23] In voltammetric methods, including linear sweep voltammetry, cyclic voltammetry, differential pulse voltammetry, square-wave voltammetry, and stripping voltammetry, measurements are performed by applying a variable potential.[24] The obtained current is proportional to the analyte concentration. In amperometry, a constant amount of potential is applied as a function of time at the working electrode. Current versus time is measured, and the change in current is related to the electroactive analyte concentration. If the analyte is not electroactive, a competitive or displacement sensor format might be preferred.[25] In these cases, a labeled analyte competes

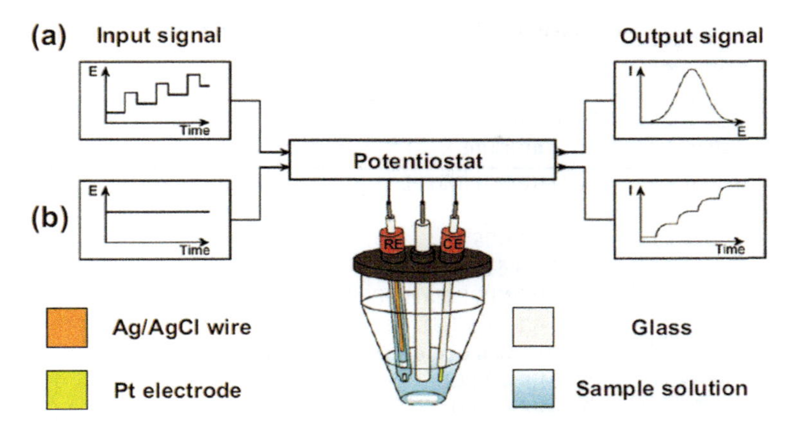

Figure 5.5 Schematic representation of (a) voltammetric and (b) amperometric sensors. RE: reference electrode; CE: counter electrode (reproduced with permission from Ref. [11]).

with the analyte for binding to the imprinted sites in the MIP or the labeled analyte first binds to the imprinted sites in the MIP and then is displaced by the analyte. The typical setup for amperometric and voltammetric sensors is given in Fig. 5.5. Additional examples of MIP-based sensors presented in the literature and their analytical performance are shown in Table 5.3.

5.2.2 MIP-Based Mass Sensors

The mass-sensitive sensor approach is getting more and more popular because it is sensitive to mass change, which is a common property for all compounds. The basic principle of mass-sensitive devices is the piezoelectric effect, which was first described by Pierre and Jacques Curie in 1880.[42] They found that when some crystalline substances are exposed to mechanical stress, they are polarized, resulting in voltage generation.[43] The inverse piezoelectric

Table 5.3 Summary of MIP-based electrochemical sensors and their performance

Analyte	Composition of membrane	Analysis method	Detection limit (M)	Ref.
Atrazine	PVC:MIP:DOP:NaTPB (25:13:55:7)	Potentiometric	5×10^{-7}	[26]
Levamisole	PVC:MIP:DBS (29.3:9.8:61)	Potentiometric	1×10^{-6}	[27]
Melamine	PVC:MIP:NPOE:NaTFBP (31.3:6.1:61.1:1.5)	Potentiometric	n/a	[28]
Atrazine	MIP in a membrane form	Potentiometric	3×10^{-5}	[29]
Oxytetracycline	PVC:MIP:*o*-NPOE	Potentiometric	2×10^{-5}	[30]
Ciprofloxacin	PVC:MIP:*o*-NPOE (35.4:2.6:61.9)	Potentiometric	1×10^{-5}	[31]
Phenylalanine	PP:octanethiol	Conductometric	3×10^{-6}	[32]
Glucose	PPD:1-dodecanethiol	Conductometric	5×10^{-5}	[33]
Glutathione	PPD:1-dodecanethiol	Conductometric	1.25×10^{-6}	[34]
Cholesterol	PMBI:1-dodecanethiol	Conductometric	4.2×10^{-7}	[35]
Theophylline	PP	Conductometric	1×10^{-6}	[36]
Morphine	MIP in an agarose gel	Amperometric	1.8×10^{-7}	[37]
Morphine	MIP in an electrosynthesized polymer	Amperometric	3×10^{-7}	[38]
Paraoxon	MIP in a poly(epichlorohydrin) matrix	Voltammetric	1×10^{-9}	[39]
Folic acid	MIP in a sol-gel matrix	Voltammetric	4.8×10^{-9}	[40]
Ephedrine	MIP in an electropolymerized polymer	Voltammetric	5×10^{-4}	[41]

PVC: poly(vinyl chloride); DOP: dioctylphthalate; NaTPB: sodium tetraphenylborate; *o*-NPOE: 2-nitrophenyloctyl ether; NaTFBP: sodium tetrakis [3,5-*bis*(trifluoro-methyl)phenyl]borate; DBS: sebacic acid dibutyl ester; PPD: poly(O-phenylen-diamine); PMBI: poly-2-mercaptobenzilimidazole; PP: polyphenol; n/a: not available.

effect discovered by Lippmann[44] describes that when an electrical field is applied to crystalline substances, mechanical deformation occurs. This inverse piezoelectric effect is used in the production and detection of sound, the generation of high voltage, electric frequency generation, and microbalances. Materials with such mechanical and electrical properties are known as piezoelectric materials, and quartz is the best-known example. The sensitivity of piezoelectric materials to changes in mass was first discovered by German physicist Sauerbrey[45] in 1959. He described that the resonance frequency of an oscillating piezoelectric material shifts due to an increase in mass as follows:

$$\Delta f \propto -f_o^2 \Delta m, \tag{5.1}$$

where f_o is the fundamental resonance frequency, Δm is the change in the deposited mass, and Δf is the frequency shift or the change in the frequency.

The quartz crystal microbalance (QCM) is one of the important applications of piezoelectric materials in that a change in the resonance frequency of the quartz crystal appears as a result of mass increase on the crystal surface. A QCM is composed of an AT-cut quartz plate coated with two gold electrodes placed one on each face, as shown in Fig. 5.6.

As represented in Fig. 5.6, the surface of one of the electrodes is coated with an MIP in film format. When an oscillating voltage is applied, the crystal oscillating at the natural resonance frequency of quartz[46] deforms. When an MIP on the surface of an electrode binds the target compound, a mass change occurs and it results in a decrease in the resonance frequency, which is proportional to the amount of target compound. It is a well-known fact that a 1 ng change in mass leads to a 1 Hz change in the resonance frequency of the quartz crystal resonating at 10 MHz.[47] QCM devices are commercially available in different resonating frequencies, between 5 MHz and

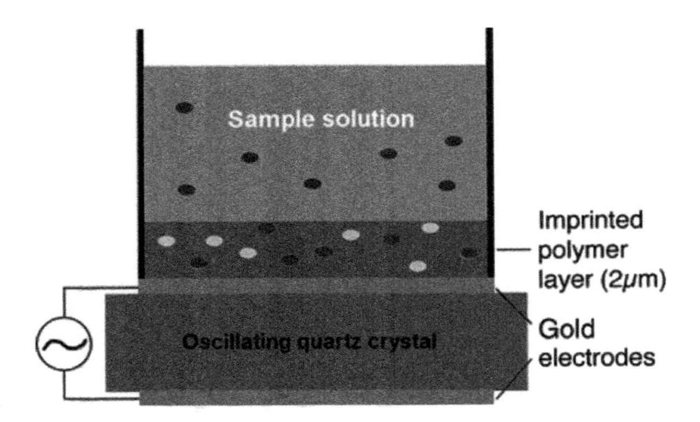

Figure 5.6 Schematic representation of a QCM (reproduced with permission from Ref. [46]).

50 MHz. The most widely used one is 10 MHz, with the sensitivity reaching 1 Hz/ng.[43] One of the first examples of an MIP-based QCM sensor was developed by Percival[48] for the determination of L-menthol in 2001.[15] Analytes both in solution[46,49–51] and in gas phase[52,53] can be determined using the QCM technique. However, while a sample in solution form is analyzed, a QCM sensor is sensitive to not only changes in mass but also changes in the viscoelasticity close to the surface.[54] In this case, the problem is eliminated by using a reference sensor coated with a nonselective polymer (nonimprinted polymer) of the same type, like an MIP. An MIP-based QCM nanosensor has been developed for the determination of tobramycin (TOB).[55] In this study, the gold surface of a QCM chip was modified using allyl mercaptan to introduce double bonds and then a TOB-imprinted poly(2-hydroxyethyl methacrylate-methacryloylamido glutamic acid) film was obtained on the gold surface. It was reported that a detection limit of 5.7×10^{-12} M was achieved. In another

study, a QCM electrode surface was coated with an *S*-propranolol–imprinted polymer and the QCM sensor was able to differentiate between the *R*- and *S*-enantiomers of the drug.[46] MIP-based QCM sensors find applications in the analysis of analytes having high molecular weights, such as proteins[56] and cells.[57]

5.2.3 MIP-Based Optical Sensors

Application of MIPs to optical sensors requires not only high affinity and selective binding sites but also sensitive optical techniques to detect the binding event.[58] Various optical detection techniques have been used together with MIPs in sensor development, including absorption (ultraviolet-visual [UV-Vis] and infrared [IR]), Raman scattering, fluorescence, and surface plasmon resonance (SPR). The change in optical properties, such as absorption, reflection, and fluorescence, of an MIP material is determined by the interaction of analyte and MIP.

Several sensors have been developed for the detection of UV-vis light–absorbing analytes. Changes in the absorbance of either the solution containing the analyte or the polymer phase are monitored. In one type of application, microtiter plates were coated with various MIPs and analysis was performed using a microtiter plate optical absorbance reader. One such study was presented by Piletsky et al.,[59] in which an MIP imprinted with epinephrine was grafted onto the surface of microtiter plates and provided a sensitivity in the 1–100 μM range. In addition, Mullet et al. reported a study related to molecularly imprinted solid-phase microextraction for the determination of propranolol in biological fluids over the concentration range of 0.5–100 μg mL^{-1} using

UV spectroscopy.[60] Vaihinger et al. determined the binding capacities of MIP nanospheres imprinted with L- or D-Boc-phenylalanine anilide.[61] Since a significant disadvantage of absorption-based sensors is the radiation dispersion at lower wavelengths, MIPs containing dyes absorbing at longer wavelengths are also used.[58] For example, Gräfe et al.[62] prepared a thin polymeric membrane containing a dye having a chromophore group that changes color upon reacting with an analyte. Absorption measurements also provide valuable information in analyzing the interaction mechanisms between a functional monomer, a cross-linker, and a template in a prepolymerization mixture.[58]

IR spectroscopy is often used to characterize structural properties of organic materials, and it has also been used to understand the interaction mechanisms between the template molecule and the functional monomer in the molecular imprinting technique.[63] In addition, Jakusch et al.[64] prepared an IR evanescent wave sensor for the detection of herbicide 2,4-dichlorophenoxyacetic acid using an MIP material immobilized onto the surface of a zinc selenide attenuated total reflection crystal. Thereby, target compound detection was achieved by monitoring the absorption signal at 1595 cm^{-1} and 1410 cm^{-1} in the mid-IR region.

Surface-enhanced Raman scattering (SERS) is becoming more and more popular because of its high sensitivity and nondestructive nature.[65] On the other hand, due to high reactivity and instability problems resulting from SERS-active surfaces, SERS has limited usage. MIPs can be used to improve the problematic properties of SERS-active surfaces by providing recognition ability and an insulating substrate surface.[65] Raman scattering produces very selective and sensitive vibrational spectra representing the characteristics of adsorbed target compounds in an imprinted polymeric layer immobilized on a metal surface.[63] Wulff et al. studied MIP layers by the SERS technique for the first time[66] and reported adsorption properties of two chiral dicarboxylic acids. Holthoff et al. synthesized an MIP xerogel via the sol-gel technique for the SERS detection of 2,4,6-trinitrotoluene.[67]

SPR is an optical technique detecting the refractive index changes that occur on thin metal films such as Au and Ag, resulting from molecular recognition or chemical transformations.[68] In SPR sensors, MIPs are used as selective recognition elements in that the polymer is coated on the surface of the metallic film and the SPR refractive index changes of the adsorbed polymer layer are monitored when the analyte interacts with the MIP.[58] Although the application of the method is mostly limited to the molecules having high molecular weights, such as peptides and proteins, several studies in combination with SPR and MIPs have been reported.[69–72]

Since fluorescence is one of the most sensitive detection techniques in comparison to other spectroscopic techniques, most of the studies on optical sensors in the field of molecular imprinting have been based on fluorescence sensing.[73] Monitoring of the binding event in MIP-based fluorescence sensors can be grouped into three main categories, which are shown in Fig. 5.7.[74]

In the first approach, since the analyte itself is fluorescent, the fluorescence of the analyte in the solution is measured before and after the equilibrium binding assay and the fluorescence intensity of the imprinted polymer is measured after the binding of fluorescent analyte. Here, the

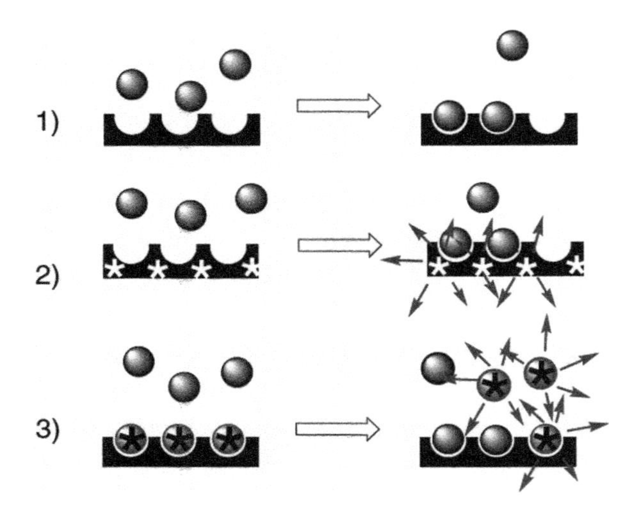

Figure 5.7 Binding events in MIP-based fluorescence sensors: (1) the analyte itself is fluorescent; (2) the analyte is nonfluorescent, and the MIP is fluorescent; (3) the analyte is nonfluorescent, and a fluorescently labeled analyte derivative or its analog is used (reproduced with permission from Ref. [74]).

role of the MIP is only to selectively recognize the analyte molecule and fluorescence signal results from the analyte and not from the MIP. Although most of analytes suffer from weak intrinsic fluorescence, leading to a limitation in the application of a fluorescent sensor for the determination of analytes, there are a number of studies in the literature using fluorescent analytes and MIPs in optical sensor design.[75–82]

In the second approach, the analyte itself is not fluorescent and the polymer functions both as a recognition element and a signaling element, reporting the binding event to the fluorescence transducer. The signaling property of the polymer may originate from a fluorophore

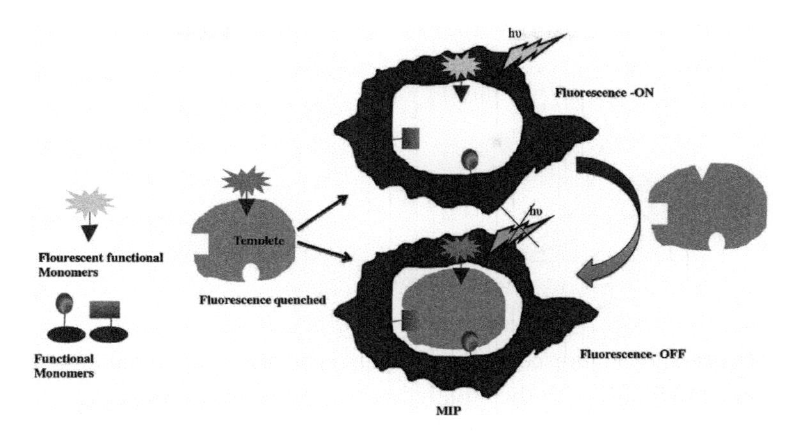

Figure 5.8 Schematic representation of the interaction of analyte with fluorescent MIP (reproduced with permission from Ref. [83]).

attached chemically to the polymer matrix or a fluorescent monomer incorporated into the polymeric chain during the synthesis of the MIP. Analyte binding is monitored via a change in the fluorescence emission intensity, the life time, or the emission wavelength.[58,73] Due to interaction and, in turn, a binding event taking place between analyte and fluorophore, fluorescence quenching appears as represented in Fig. 5.8.[83] The extent of quenching should be directly proportional to the amount of the bound target analyte. Various studies based on the quenching mechanism have been reported in the literature.[84–87]

However, a high background signal originating from the imprinted polymer causes the sensitivity of the fluorescent sensor to decrease. This limitation can be eliminated by designing and developing new fluorescent monomers that turn on upon binding.[74] In this case, the fluorescence signal is enhanced when the analyte binds, and the increase in

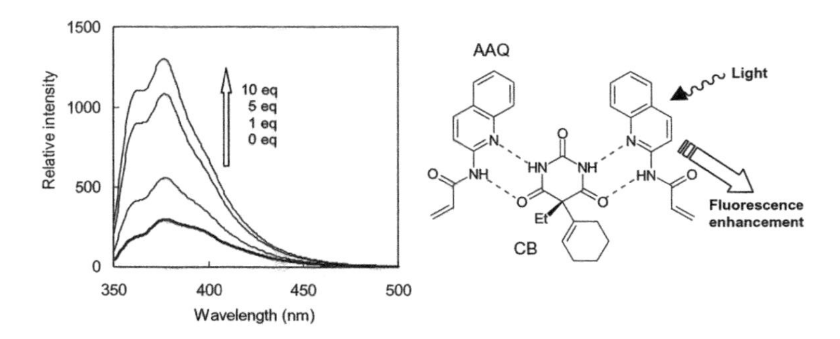

Figure 5.9 Fluorescence spectral changes for 2-acrylamidoquinoline (12.5 μM) with the addition of various concentrations (0, 1, 5, and 10 equiv.) of cyclobarbital to $CDCl_3$ at $\lambda_{ex} = 330$ nm (reproduced with permission from Ref. [88]).

the fluorescence intensity becomes directly proportional to the amount of the bound analyte. For example, Takeuchi et al. synthesized a cyclobarbital-imprinted MIP containing fluorescent monomer 2-acrylamidoquinoline.[88] This monomer was designed and then synthesized to enhance the fluorescence intensity of the MIP when cyclobarbital was bound to the 2-acrylamidoquinoline situated in the vicinity of the imprinted cavity via hydrogen bonding. As shown in Fig. 5.9, the fluorescence intensity of the MIP was enhanced when various amounts of cyclobarbital were bound to the 2-acrlyamidoquinoline functional monomer. The enhancement in fluorescence can be easily attributed to the increased rigidity of 2-acrylamidoquinoline due to the formation of multiple hydrogen bonding.

On the other hand, if the analyte does not interact with the fluorophore or the fluorescent monomer inside the polymeric matrix strongly enough, resulting in insufficient change in fluorescence properties (fluorescence intensity, life time, or emission wavelength), an external quencher or modifier might be used, through a displacement mechanism, to increase the fluorescence change upon analyte binding.[73]

Another popular fluorescent sensor application is based on the combination of MIPs and quantum dots (QDs). In this type of fluorescence sensor, the MIPs and the QDs function as a recognition element and a signaling element, respectively, reporting the analyte binding to the transducer. The QD in nanodimensions and the MIP obtained through the surface imprinting approach form the core and the shell of the particle, respectively. The fluorescence signal of the MIP-coated QDs might be quenched or enhanced when the analyte molecules are bound to the recognition cavities of the MIP shell.[89,90]

In the third approach, the analyte is nonfluorescent and a fluorescently labeled analyte derivative or its analog is used as a signaling element. In this case, the MIP is also nonfluorescent and takes part as a recognition element in the fluorescent sensor development. The binding event is monitored via a competitive or displacement assay.[91–93]

References

1. Kryscio, D. R. and Peppas, N. A. (2012). Crictical review and perspective of macromolecularly imprinted polymers, *Acta Biomater.*, **8**, pp. 461–473.

2. Perez-Moral, N. and Mayes, A. G. (2004). Comparative study of imprinted polymer particles prepared by different polymerisation methods, *Anal. Chim. Acta*, **504**, pp. 15–21.

3. Yan, C., Lu, Y. and Gao, S. (2009). Preparation and characterization of PS/pAPBA core-shell microspheres, *Front. Chem. China*, **4**(2), pp. 168–172.

4. Mayes, A. G. and Whitcombe, M. J. (2005). Synthetic strategies for the generation of molecularly imprinted organic polymers, *Adv. Drug Deliv. Rev.*, **57**, pp. 1742–1778.
5. Wulff, G. and Sarhan, A. (1972). The use of polymers with enzyme-analogous structures for the resolution of racemates, *Angew. Chem. Int. Ed.*, **11**, p. 4.
6. Kandimalla, V. B. and Ju, H. (2004). Molecular imprinting: a dynamic technique for diverse applications in analytical chemistry, *Anal. Bioanal. Chem.*, **380**, pp. 587–605.
7. Arshady, R. and Mosbach, K. (1981). Synthesis of substrate-selective polymers by host-guest polymerization, *Makromol. Chem.*, **182**, pp. 687–692.
8. Pichon, V. and Chapuis-Hugon, F. (2008). Role of molecularly imprinted polymers for selective determination of environmental pollutants: a review, *Anal. Chim. Acta*, **622**, pp. 48–61.
9. Alexander, C., Andersson, H. S., Andersson, L. I., Ansell, R. J., Kirsch, N., Nicholls, I. A., O'Mahony, J. and Whitcombe, M. J. (2003). Molecular imprinting science and technology: a survey of the literature for the years up to and including 2003, *J. Mol. Recognit.*, **19**, pp. 106–180.
10. Holthoff, E. L. and Bright, F. V. (2007). Molecularly templated materials in chemical sensing, *Anal. Chim. Acta*, **594**, pp. 147–161.
11. Li, S., Ge, Y., Piletsky, S. A. and Lunec, J. (2012). *Molecularly Imprinted Sensors: Overview and Applications*, Antuna-Jimenez, D., Diaz-Diaz, G., Blanco-Lopez, M. C., Lobo-Castanon, M. J., Mirando-Ordieres, A. J. and Tunon-Blanco, P., Chapter 1: Molecularly imprinted electrochemical sensors: past, present and future, Elsevier, UK, pp. 1–33.
12. Yan, M. and Ramström, O. (2005). *Molecularly Imprinted Materials Science and Technology*, Haupt, K., Chapter 26: Molecularly imprinted polymers as recognition elements in sensors: mass and electrochemical sensors, Marcel Dekker, New York, USA, pp. 685–700.

13. Boyle, A., Genies, E. M. and Lapkowski, M. (1989), Application of electronic conducting polymers as sensors-polyaniline in the solid-state for detection of solvent vapors and polypyrrole for detection of biological ions in solutions, *Synth. Met.*, **28**, pp. 769–774.

14. Vinokurov, I. A. (1992). A new kind of redox sensor based on conducting polymer films, *Sens. Actuators, B*, **10**, pp. 31–35.

15. Li, S., Ge, Y., Piletsky, S. A. and Lunec, J. (2012). *Molecularly Imprinted Sensors: Overview and Applications*, Piletsky, S., Piletsky, S. and Chianella, I., Chapter 14: MIP-based sensors, Elsevier, UK, pp. 339–354.

16. Piletsky, S. A., Kurys, Y. I, Rachkov A. E. and Elskaya, A. V. (1994). Formation of matrix polymers sensitive to aniline and phenol, *Russ. J. Electrochem.*, **30**, pp. 1090–1092.

17. Hedborg, E., Winquist, F., Lundström, I., Andersson, L. I. and Mosbach, K. (1993). Some studies of molecularly imprinted polymers membranes in combination with field-effect devices, *Sens. Actuators, A*, **37**, pp. 796–799.

18. Prasada Rao, T. and Kala, R. (2008). Formation of matrix polymers sensitive to aniline and phenol, *Talanta*, **76**, pp. 485–496.

19. Kriz, D., Kempe, M. and Mosbach, K. (1996). Introduction of molecularly imprinted polymers as recognition elements in conductometric chemical sensors, *Sens. Actuators, B*, **33**, pp. 178–181.

20. Piletsky, S. A., Piletskaya, E. V., Elgersma, A. V., Yano, K., Karube, I., Parhometz, Y. P. and El'skaya, A. V. (1995). Atrazine sensing by molecularly imprinted, *Biosens. Bioelectron.*, **10**, pp. 959–964.

21. Suedee, R., Intakong, W. and Dickert, F. L. (2006). Molecularly imprinted polymer-modified electrode for on-line conductometric monitoring of haloacetic acids in chlorinated water, *Anal. Chim. Acta*, **569**, pp. 66–75.

22. Suedee, R., Srichana, T., Sangpagai, C., Tunthana, C. and Vanichapichat, P. (2004). Development of trichloroacetic acid

sensor based on molecularly imprinted polymer membrane for the screening of complex mixture of haloacetic acids in drinking water, *Anal. Chim. Acta*, **504**, pp. 89–100.

23. Kissinger, P. T. and Heineman, W. R. (1996). *Laboratory Techniques in Electroanalytical Chemistry*, Marcel Dekker Inc., New York, USA.
24. Ronkainen, N. J., Halsal H. B. and Heineman W. R. (2010). Electrochemical biosensors, *Chem. Soc. Rev.*, **39**, pp. 1747–1763.
25. Kriz, D. and Mosbach, K. (1995). Competitive amperometric morphine sensor based on an agarose immobilized molecularly imprinted polymer, *Anal. Chim. Acta*, **300**, pp. 71–75.
26. Prasad, K., Prathish, K. P., Gladis, J. M., Naidu, G. R. K. and Rao, T. P. (2007). Molecularly imprinted polymer (biomimetic) based potentiometric sensor for atrazine, *Sens. Actuators, B*, **123**, pp. 65–70.
27. Sadeghi, S., Fathi, F. and Abbasifar, J. (2007). Potentiometric sensing of levamisole hydrochloride based on molecularly imprinted polymer, *Sens. Actuators, B*, **122**, pp. 158–164.
28. Liang, R., Zhang, R. and Qin, W. (2009). Potentiometric sensor based on molecularly imprinted polymer for determination of melamine in milk, *Sens. Actuators, B*, **141**, pp. 544–550.
29. D'Agostino, G., Alberti, G., Biesuz, R. and Pesavento, M. (2006). Potentiometric sensor for atrazine based on a molecular imprinted membrane, *Biosens. Bioelectron.*, **22**, pp. 145–152.
30. Hutchins, R.S. and Bachas, L.G. (1995). Nitrate-selective electrode developed by electrochemically mediated imprinting/doping of polypyrrole. *Anal. Chem.*, **67**, pp. 1654–1660.
31. Oliveria, H. M. V., Moreira, F. T. C. and Sales, M. G. F. (2011). Ciprofloxacin-imprinted polymeric receptors as ionophores for potentiometric transduction, *Electrochim. Acta*, **56**, pp. 2017–2023.
32. Panasyuk, T. L., Mirsky, V. M., Piletsky, S. A. and Wolfbeis, O. S. (1999). Electropolymerized molecularly imprinted polymers as

receptor layers in capacitive chemical sensors, *Anal. Chem.*, **71**, pp. 4609–4613.

33. Cheng, Z. L., Wang, E. K. and Yang, X. R. (2001). Capacitive detection of glucose using molecularly imprinted polymers, *Biosens. Bioelectron.*, **16**, pp. 179–185.

34. Yang, L., Wei, W., Xia, J., Tao, H. and Yang, P. (2005). Capacitive biosensor for glutathione detection based on electropolymerized molecularly imprinted polymer and kinetic investigation of the recognition process, *Electroanalysis*, **17**, pp. 969–977.

35. Aghaei, A., Hosseini, M. R. M. and Najafi, M. (2010). A novel capacitive biosensor for cholesterol assay that uses an electropolymerized molecularly imprinted polymer, *Electrochim. Acta*, **55**, pp. 1503–1508.

36. Wang, Z., Kang, J., Liu, X. and Ma, Y. (2007). Capacitive detection of theophylline based on electropolymerized molecularly imprinted polymer, *Int. J. Polym. Anal. Charact.*, **12**, pp. 131–142.

37. Kriz, D. and Mosbach, K. (1995). Competitive amperometric morphine sensor based on an agarose immobilised molecularly imprinted polymer, *Anal. Chim. Acta*, **300**, pp. 71–75.

38. Ho, K. C., Yeh, W. M., Tung, T. S. and Liao, J. Y. (2005). Amperometric detection of morphine based on poly (3,4-ethylenedioxythiophene) immobilized molecularly imprinted polymer particles prepared by precipitation polymerization, *Anal. Chim. Acta*, **542**, pp. 90–96.

39. Alizadeh, T. (2010). Comparison of different methodologies for integration of molecularly imprinted polymer and electrochemical transducer in order to develop a paraoxon voltammetric sensor, *Thin Solid Films*, **518**, pp. 6099–6106.

40. Prasad, B. B., Madhuri, R., Tiwari, M. P. and Sharma, P. S. (2010). Electrochemical sensor for folic acid based on a hyperbranched molecularly imprinted polymer-immobilized sol–gel-modified pencil graphite electrode, *Sens. Actuators, B*, **146**, pp. 321–330.

41. Mazzotta, E., Picca, R. A., Malitesta, C., Piletsky, S. A. and Piletska, E. V. (2008). Development of a sensor prepared by

entrapment of MIP particles in electrosynthesised polymer films for electrochemical detection of ephedrine, *Biosens. Bioelectron.*, **23**, pp. 1152–1156.

42. Curie, J. and Curie, P. (1880). Contractions et dilatations produites par des tensions électriques dans les cristaux hémièdres à faces inclinées, *C.R. Acad. Sci.*, **93**, pp. 1137–1140.
43. Li, S., Ge, Y., Piletsky, S. A. and Lunec, J. (2012) *Molecularly Imprinted Sensors: Overview and Applications*, Mujahid, A. and Dickert, F. L., Chapter 6: Molecularly imprinted polymers for sensors: comparison of optical and mass-sensitive detection, Elsevier, UK, pp. 125–159.
44. *Lippmann, G. (1881).* Principe de la conservation de l'électricité [Principle of the conservation of electricity], *Annal. Chim. Phys. (in French)*, **24**, p. 145.
45. Sauerbrey, G. (1959). Verwendung von Schwingquarzen zur Wägung dünner Schichten und zur Mikrowägung, *Z. Phys. Chem.*, **155**, pp. 206–222.
46. Haupt, K., Noworyta, K. and Kutner, W. (1999). Imprented polymer based enantioselective acoustic sensor using a quartz crystal microbalance, *Anal. Commun.*, **36**, pp. 391–393.
47. Hillberg, A. L., Brain, K. R. and Allender, C. J. (2005). Molecular imprinted polymer 'sensors: implications for therapeutics, *Adv. Drug Deliv. Rev.*, **57**, pp. 1875–1889.
48. Percival, C. J., Stanley, S., Galle, M., Braithwaite, A., Newton, M. I., McHale, G. and Hayes, W. (2001). molecular-imprinted, polymer-coated quartz crystal microbalances for the detection of terpenes, *Anal. Chem.*, **73**, p. 4225.
49. Dickert, F. L, Tortschanoff, M., Bulst, W. E. and Fischerauer, G. (1999). Molecularly imprinted sensor layers for the detection of polycyclic aromatic hydrocarbons in water, *Anal. Chem.*, **71**, pp. 4559–4563.
50. Malitesta, C., Losito, I. and Zambonin, P. G. (1999). Molecularly imprinted electrosynthesized polymers: new materials for biomimetic sensors, *Anal. Chem.*, **71**, pp. 1366–1370.

51. Liang, C., Peng, H., Bao, X., Nie, L. and Yao, S. (1999). Study of molecular imprinting polymer coated baw bio-mimetic sensor and its application to the determination of caffeine in human serum and urine, *Analyst*, **124**, pp. 1781–1785.

52. Dickert, F. L., Fort, P., Lieberzeit, P. and Tortschanoff, M. (1998). Molecularly imprinted in chemical sensing-detection of aromatic and halogenated hydrocarbons as well as polar solvent vapors, *Fresenius J. Anal. Chem.*, **360**, pp. 759–762.

53. Ji, H. S., McNiven, S., Ikebukuro, K. and Karube, I. (1999). Selective piezoelectric odor sensors using molecularly imprinted polymers, *Anal. Chem. Acta*, **390**, pp. 93–100.

54. Yan, M. and Ramström, O. (2005). *Molecularly Imprinted Materials Science and Technology*, Haupt, K., Chapter 26: Molecularly imprinted polymers as recognition elements in sensors: mass and electrochemical sensors, Marcel Dekker, New York, USA, pp. 685–700.

55. Yola, M. L., Uzun, L., Özaltın, N. and Denizli, A. (2014). Development of molecular imprinted nanosensor for determination of tobramycin pharmaceuticals and foods, *Talanta*, **120**, pp. 318–324.

56. Tai, D. F., Lin, C. Y., Wu, T. Z., Huang, J. H. and Shu, P. Y. (2006). Artificial receptors in serologic tests for the early diagnosis of dengue virus infection, *Clin. Chem.*, **52**, pp. 1486–1491.

57. Dickert, F. L. and Hayden, O. (2002), Bioimprinting of polymers and sol–gel phases. Selective detection of yeasts with imprinted polymers, *Anal. Chem.*, **74**, pp. 1302–1306.

58. Moreno-Bondi, M. C., Navarro-Villoslada, F., Benito-Pena, E. and Urraca, J. L. (2008). Molecularly imprinted polymers as selective recognition elements in optical sensing, *Curr. Anal. Chem.*, **4**, pp. 316–340.

59. Piletsky, S. A., Piletska, E. V., Chen, B. N., Karim, K., Weston, D., Barett, G., Lowe, P. and Turner, A. P. F. (2000). Chemical grafting of molecularly imprinted homopolymers to the surface of microplates. Application of artificial adrenergic receptor in

enzyme-linked assay for β-Agonists determination, *Anal. Chem.*, **72**, pp. 4381–4385.

60. Mullet, W. M., Martin, P. and Pawliszyn, J. (2001). In-tube molecularly imprinted polymer solid-phase microextraction for the selective determination of propranolol, *Anal. Chem.*, **73**, pp. 2383–2389.
61. Vaihinger, D., Landfester, K., Kräuter, I., Brunner, H. and Tovar, G. E. M. (2002). Molecularly imprinted polymer nanospheres as synthetic affinity receptors obtained by miniemulsion polymerization, *Macromol. Chem. Phys.*, **203**, pp. 1965–1973.
62. Gräfe, A., Haupt, K. and Mohr, G. J. (2006). Optical sensor materials for the detection of amines in organic solvents, *Anal. Chim. Acta*, **565**, pp. 42–47.
63. Henry, O. Y., Cullen, D. C. and Piletsky, S. A. (2005). Optical interrogation of molecularly imprinted polymers and development of MIP sensors: a review, *Anal. Bioanal. Chem.*, **382**, pp. 947–956.
64. Jakusch, M., Janotta, M. and Mizaikoff, B. (1999). Molecularly imprinted polymers and infrared evanescent wave spectroscopy. A chemical sensors approach, *Anal. Chem.*, **71**, pp. 4786–4791.
65. Uzun, L. and Turner, A. P. F. (2016). Molecularly-imprinted polymer sensors: realising their potential, *Biosens. Bioelectron.*, **76**, pp. 131–144.
66. Kostrewa, S., Emgenbroich, M., Klockow, D. and Wulff, G. (2003). Surface-enhanced raman scattering on molecularly imprinted polymers in water, *Macromol. Chem. Phys.*, **204**, pp. 481–487.
67. Holthoff, E. L., Cullum, D. N. and Hankus, M. E. (2011). A nanosensor for TNT detection based on molecularly imprinted polymers and surface enhanced Raman scattering, *Sensors*, **11**, pp. 2700–2714.
68. Frasconi, M., Tel-Vered, R., Riskin, M. and Willner, I. (2010). Surface plasmon resonance analysis of antibiotics using imprinted boronic acid-functionalized au nanoparticle composites, *Anal. Chem.*, **82**, pp. 2512–2519.

69. Lai, E. P. C., Fafara, A., VanderNoot, V. A., Kono, M. and Polsky, B. (1998). Surface plasmon resonance sensors using molecularly imprinted polymers for sorben assay of theophylline caffeine and xanthine, *Can. J. Chem.*, **76**, pp. 265–273.

70. Kugimiya, A. and Takeuchi, T. (2001). Surface plasmon resonance sensor using molecularly imprinted polymer for detection of sialic acid, *Biosens. Bioelectron.*, **16**, pp. 1059–1062.

71. Yola, M. L., Atar, N. and Eren T. (2014). Determination of amikacin in human plasma by molecular imprinted SPR nanosensor, *Sens. Actuator, B*, **198**, pp. 70–76.

72. Yola, M. L., Atar, N. and Erdem A. (2015). Oxytocin imprinted polymer based surface plasmon resonance sensor and its application to milk sample, *Sens. Actuators, B*, **221**, pp. 842–848.

73. Yan, M. and Ramström, O. (2005). *Molecularly Imprinted Materials Science and Technology*, Gao, S., Wang, W. and Wang, B., Chapter 27: Molecularly imprinted polymers as recognition elements in optical sensors, Marcel Dekker, New York, USA, pp. 701–726.

74. Stephenson, C. J. and Shimizu, K. D. (2007). Colorimetric and fluometric molecularly imprinted polymer sensors and binding assays, *Poly. Int.*, **56**, pp. 482–488.

75. Kriz, D., Ramström, O., Svensson, A. and Mosbach, K. (1995). Introducing biomimetic sensors based on molecularly imprinted polymers as recognition elements, *Anal. Chem.*, **67**, 2142–2144.

76. Matsui, J., Kubo, H. and Takeuchi, T. (2000). Molecualrly imprinted fluorescent-shift receptors prepared with 2-(trifluoromethyl) acrylic acid, *Anal. Chem.*, **72**, pp. 3286–3290.

77. Suarez-Rodriguez, J. L. and Diaz Garcia, M. E. (2000). Flavonol fluorescent flow-through sensing based on a molecular imprinted polymer, *Anal. Chim. Acta*, **405**, pp. 67–76.

78. Rachkov, A., McNiven, S., El'skaya, A., Yano, K. and Karube, I. (2000). Fluorescent detection of β-estradiol using a molecularly imprinted polymer, *Anal. Chim. Acta*, **405**, pp. 23–29.

79. Lulka, M. F., Chambers, J.B., Valdes, E. R., Thompson, R. G. and Valdes, J. J. (1997). Molecular imprinting of small molecules with organic silanes: fluorescence detection, *Anal. Lett.*, **30**, pp. 2301–2313.

80. Dickert, F. L. and Thierer, S. (1996). Molecularly imprinted polymers for optiochemical sensors, *Adv. Mater.*, **8**, pp. 987–990.

81. Dickert, F. L., Besenbock, H. and Tortschanoff, M. (1998). Molecular imprinting through van der Walls interactions: fluorescence detection of pH as water, *Adv. Mater.*, **10**, pp. 149–151.

82. Dickert, F. L. and Tortschanoff, M. (1999). Molecularly imprinting sensor layers for the detection of polycylic aromatic hydrocarbons in water, *Anal. Chem.*, **71**, pp. 4559–4563.

83. Rouhani, S. and Nahavandifard, F. (2014). Molecularly imprinting-based fluorescent optosensor using a polymerizable 1,8-naphthalamide dye as a fluorescence functional monomer, *Sens. Actuators, B*, **197**, pp. 185–192.

84. Wang, W., Gao, S. and Wang, B. (1999). Building fluorescent sensors by template polymerization: the preparation of a fluorescent sensor for D-fructose, *Org. Lett.*, **1**, pp. 1209–1212.

85. Gao, S., Wang, W. and Wang, B. (2001). Building fluorescent sensors for carbohydrates using template-directed polymerizations, *Bioorg. Chem.*, **29**, pp. 308–320.

86. Turkewitsch, P., Wandelt, B., Darling, G. D. and Powell, W. S. (1998). Fluorescent functional recognition sites through molecular imprinting. A polymer-based fluorescent chemosensor for aqueous cAMP, *Anal. Chem.*, **70**, pp. 2025–2030.

87. Zhang, H., Verboom, W. and Reinhoudt, D. N. (2001). 9-(Guanidinomethyl)-10-vinylanthracene: a suitable fluorescent monomer for MIPs, *Tetrahed. Lett.*, **42**, pp. 4413–4416.

88. Kubo, H., Yoshioka, N. and Takeuchi, T. (2005). Fluorescent imprinted polymers prepared with 2-acrylamidoquinoline as a signaling monomer, *Org. Lett.*, **7**, pp. 359–362.

89. Lin, C. İ., Joseph, A. K., Chang, C. K. and Lee, Y. D. (2004). Synthesis and photoluminescence study of molecularly imprinted poly-

mers appended onto CdSe/ZnS core-shells, *Biosens. Bioelectron.*, **20**, pp. 127–131.

90. Zhang, W. E., He, X. W., Chen, Y., Li, W. Y. and Zhang, Y. K. (2011). Composite of CdTe quantum dots and molecularly imprinted polymer as a sensing material for cytochrome C, *Biosens. Bioelectron.*, **26**, pp. 2553–2558.

91. Haupt, K., Mayes, A. G. and Mosbach, K. (1998). Herbicide assay using an imprinted polymer-based system analogous to competitive fluoroimmunoassays, *Anal. Chem.*, **70**, pp. 3936–3939.

92. Piletsky, S. A., Terpetschnig, E., Andersson, H. S., Nicholls, I. A. and Wolfbeis, O. S. (1999). Application of non-specific fluorescent dyes for monitoring enantio-selective ligand binding to molecularly imprinted polymers, *Fresenius J. Anal. Chem.*, **364**, pp. 512–516.

93. Levi, R., McNiven, S., Piletsky, S. A., Cheng, S. H., Yano, K. and Karube, I. (1997). Optical detection of chloramphenicol using molecularly imprinted polymers, *Anal. Chem.*, **69**, pp. 2017–2021.

Index

9789814800891